Lecture Notes in Mathematics

Edited by A. Dold and B. Eckmann

491

Kevin A. Broughan

Invariants for Real-Generated Uniform Topological and Algebraic Categories

Springer-Verlag
Berlin · Heidelberg · New York 1975

Author
Kevin A. Broughan
Department of Mathematics
University of Waikato
Hamilton/New Zealand

Library of Congress Cataloging in Publication Data

Broughan, K A 1943-
Invariant for real-generated uniform and algebraic categories.

(Lecture notes in mathematics ; 491)
Bibliography: p.
Includes index.
1. Uniform spaces. 2. Topological spaces.
3. Categories (Mathematics) 4. Invariants.
I. Title. II. Series: Lecture notes in mathematics (Berlin) ; 491.
QA3.L28 no. 491 [QA611.25] 510'.8s [514'.32]
75-34130

AMS Subject Classifications (1970): 02E15, 02E99, 06A10, 10A40, 10B05, 10F35, 10M10, 12J10, 12J20, 13A15, 13J99, 15A03, 18B99, 18D35, 22A99, 26A51, 33A70, 46A15, 54C05, 54C30, 54D20, 54E05, 54E15, 54E25, 54E35, 54F45, 54F50, 54H25.

ISBN 3-540-07418-X Springer-Verlag Berlin · Heidelberg · New York
ISBN 0-387-07418-X Springer-Verlag New York · Heidelberg · Berlin

Printed in Germany

Offsetdruck: Julius Beltz, Hemsbach/Bergstr.

Introduction

This book is concerned with the construction of natural invariants for categories and algebraic categories of uniform and topological spaces. In each of these categories the natural invariants are similar in definition and form.

The invariants are constructed from the set of values of real valued functions which have generated the structures in each category. An illustration of the method of construction of the invariants follows -- this is the construction in the category of metrizable topological spaces. If a topological space has a compatible metric having its range in a certain subset of the real numbers, then every homeomorphic space has also such a compatible metric. Hence, for each subset of the positive real numbers there will be, at least theoretically, a corresponding metrization theorem, which depends on the structure of the subset chosen. However, not all subsets give rise to unique metrization theorems. If two subsets give rise to the same metrization theorem they are said to be equivalent. More precisely, two subsets of the positive real numbers are called equivalent if each metrizable topological space with a compatible metric having its range in one of the subsets has a compatible metric with range in the other subset, and vice versa. Each equivalence class formed from this relation is a distinct invariant. The set of all equivalence classes thus induced has been given a partial order. One of the objects of this book is to characterize each distinct equivalence class in topological terms. However it is not necessary to deal with the class which is represented by the set of positive real numbers themselves since this characterization is provided by the metrization theorem of Nagata, Smirnov and Bing.

The following is a brief summary of the content of the book.

In Chapter I, the notions of a real-generated category and of an invariant on a category, are defined. The classes of subsets of real numbers are introduced and several theorems are proved about real-generated categories and the related classes. The idea of a derived category is introduced. This notion leads to a method whereby the partially ordered sets of classes of subsets formed from different categories may be related. (This method is used frequently in this book.) Subsets of real numbers are considered to be objects in a category and several morphisms between such objects are constructed. These morphisms are very important for the development of the later theory.

In Chapter II, the special case of uniform spaces is studied. An example of the theorems proved is the following: If a metrizable uniform space has a compatible metric with range in the computable numbers, then the space has a compatible metric with range in the dyadic rationals. It is also proved that the induced partially ordered set of classes of real subsets has an initial segment consisting of a four element chain. Each of the corresponding four equivalence classes is characterized in terms of uniform structures. The class represented by the positive rational numbers is of particular interest. Finally, the theorems for metrizable uniform spaces are extended to arbitrary uniform spaces by using families of pseudometrics.

Invariants for metrizable topological and other types of topological spaces are studied in Chapter III. The following is a typical result: If a space has a compatible metric with range in a closed subset of the real numbers which is not a neighborhood of zero in the positive real numbers, then the space has large inductive dimension zero. This theorem extends earlier results. Then the category of metrizable topological spaces in considered, and the induced partially ordered set of classes of real subsets is shown to have an initial segment consisting of a three element chain. It is not yet known whether the class represented by the positive rational numbers is equal to the third member of this chain. The class of topological spaces having dimension zero is studied in some detail. It is proved that a paracompact space has large inductive dimension zero, if and only if, it has small inductive dimension zero and if every closed-open cover of the space has a closed-open locally finite refinement. Finally, the invariants are studied for classes of topological spaces whose topologies are generated by single real valued functions which satisfy various combinations of axioms, and for spaces whose topologies are generated by families of real valued functions.

In Chapter IV, induced classes are described for topological groups, rings, fields and general topological vector spaces. As the structure of the domain object becomes more complex, the structure generating functions are required to satisfy more stringent requirements -- sufficiently stringent, that is, to ensure that the algebraic operations are continuous. The theorems at each more highly structured level usually depend upon the earlier, more general, results.

In Chapter V, dense subspaces, completions, and the computable generation of spaces are studied. Particular emphasis is given to the classes of spaces introduced in the earlier chapters. Results such as the following are proved: Given any countable subset in any metric space, there exists a metric on the space, uniformly equivalent to the original metric, which has the property that the distance between any two points in the countable subset is rational. It is also shown how any complete separable metric space without isolated points, can be exhibited as the completion of the rational numbers equipped with a rational valued metric compatible with the usual topology. A form of connectedness for uniform spaces is defined and the following theorem proved: A uniform space is connected, if and only if, its completion is connected. A fixed point theorem for functions on the rational numbers is proved and the irrationality of the number "e" is demonstrated.

An index listing the symbols, categories and classes of subsets which are used throughout is included at the end.

My thanks go to Professor Edgar R. Lorch of Columbia University for his advice and encouragement during the preparation of this book.

Contents

Chapter I : Classes of subsets as invariants

Real-generated categories

Definition 1.1 Let $\mathcal{A}$ be a category. Suppose that (i) Each object X in $\mathcal{A}$ has an underlying set X, an algebraic structure $\mathfrak{D}$ which may be empty, and an additional structure $\mathfrak{A}$. We frequently write $(X,\mathfrak{D},\mathfrak{A})$ instead of X. And

(ii) The additional structure $\mathfrak{A}$ is generated according to a rule $\mathcal{R}$ by a family of real valued functions $F: (X,\mathfrak{D})^n \to R$, (where n is 1 or 2), which satisfies a given set of properties $\mathcal{P}$. When this is so we frequently write $\mathfrak{A}_F$ instead of $\mathfrak{A}$. The rule $\mathcal{R}$, the properties $\mathcal{P}$, and the number n are fixed for the given category. And

(iii) The rule $\mathcal{R}$ and the properties $\mathcal{P}$ satisfy the following condition: if $h: (X_\alpha,\mathfrak{D}_\alpha,\mathfrak{A}_\alpha) \to (X_\beta,\mathfrak{D}_\beta,\mathfrak{A}_\beta)$ is an equivalence in $\mathcal{A}$ and $F: (X_\beta,\mathfrak{D}_\beta)^n \to R$ satisfies $\mathcal{P}$ and generates $\mathfrak{A}_\beta$, then, if $G = \{f \circ h^n : f \in F\}$, the family $G: (X_\alpha,\mathfrak{D}_\alpha)^n \to R$ satisfies $\mathcal{P}$ and generates the structure $\mathfrak{A}_\alpha$. When this is so we say the properties $\mathcal{P}$ are natural

<u>for the category</u> $\mathcal{A}$ (<u>Definition 1.2</u>).

If conditions (i), (ii) and (iii) are satisfied we say $\mathcal{A}$ <u>is a real-generated category with datum</u> $V = (\mathcal{A}, \mathcal{P}, \mathcal{R})$.

Let $\mathcal{J}$ be the family of non empty subsets of R and let $\mathcal{S} \subset \mathcal{J}$ be a non empty subfamily. Let S and T be in $\mathcal{S}$. We say <u>$S \prec T$</u>, if whenever $(X, \mathfrak{D}, \mathfrak{A})$ is in $\mathcal{A}$, and there exists $F: X^n \to S$ satisfying $\mathcal{P}$ and generating $\mathfrak{A}$ according to the rule $\mathcal{R}$, then there exists a family $G: X^n \to T$ satisfying $\mathcal{P}$ and generating $\mathfrak{A}$ according to the rule $\mathcal{R}$. We say $S \sim T$ if $S \prec T$ and $T \prec S$ and then $\sim$ is an equivalence relation. If $S \in \mathcal{S}$ we denote the equivalence class to which S belongs by $[S, V, \mathcal{S}]$ where V is the datum of the category $\mathcal{A}$.

There are many examples of such categories:

(i) The category of metric spaces and isometries is real-generated by metrics. We will denote this category by $\mathcal{M}$ and write $(X, \mathfrak{M}_d)$ instead of (X,d) when d is a metric on the set X.

(ii) The categories metrizable uniform spaces $(\mathcal{U})$, metrizable topological spaces $(\mathcal{T})$, metrizable proximity spaces $(\mathcal{N})$, and metrizable convergence spaces $(\mathcal{L})$, are real-generated.

(iii) The categories uniform spaces ($\mathcal{E}$) and quasi uniform spaces ($\mathcal{K}$) are also real-generated. Each of these categories has empty algebraic structure. They will be studied especially in the first three chapters.

The order $<$ defined above induces a partial order $[S,V,\mathcal{S}] < [T,V,\mathcal{S}]$ on the set of classes $\{[S,V,\mathcal{S}] : S \in \mathcal{S}\}$ and we denote the partially ordered set so generated by $P(V,\mathcal{S})$. Frequently we will write $[S]$ instead of $[S,V,\mathcal{S}]$. Other conventions will be introduced as we proceed. Note that $S < T$ if and only if $[S,V,\mathcal{S}] < [T,V,\mathcal{S}]$.

Derived Categories

Definition 1.3 Let $\mathcal{A}$ be a real-generated category with datum $V_{\mathcal{A}} = (\mathcal{A}, \mathcal{P}_{\mathcal{A}}, \mathcal{R}_{\mathcal{A}})$ and let $\mathcal{B}$ be a real generated category with datum $V_{\mathcal{B}} = (\mathcal{B}, \mathcal{P}_{\mathcal{B}}, \mathcal{R}_{\mathcal{B}})$. We say $\mathcal{B}$ is derived from $\mathcal{A}$ if $\mathcal{P}_{\mathcal{A}} = \mathcal{P}_{\mathcal{B}}$ and the rule

$$(X,\mathfrak{D},\mathfrak{A}_F) \mapsto (X,\mathfrak{D},\mathfrak{B}_F)$$
$$f \mapsto f$$

defines a functor $\mathcal{A} \to \mathcal{B}$, which is surjective on objects where $(X,\mathfrak{D},\mathfrak{A})$ is in $\mathcal{A}$, $F\colon X^n \to R$ satisfies $\mathcal{P}_{\mathcal{A}}$ and f is a morphism in $\mathcal{A}$. It follows that if $(X,\mathfrak{D},\mathfrak{A}_F) = (X,\mathfrak{D},\mathfrak{A}_G)$ in $\mathcal{A}$, and $\mathcal{B}$ is a category derived from $\mathcal{A}$, then $(X,\mathfrak{D},\mathfrak{B}_F) = (X,\mathfrak{D},\mathfrak{B}_G)$ in $\mathcal{B}$.

<u>Definition 1.4</u> Let $\mathcal{A}$ be a real-generated category with datum $V = (\mathcal{A},\mathcal{P},\mathcal{R})$, let $(X,\mathfrak{D},\mathfrak{A})$ be in $\mathcal{A}$, and let $S \in \mathcal{S} \subset \mathcal{I}$. We say $(X,\mathfrak{D},\mathfrak{A})$ is $[S,V,\mathcal{S}]$- metrizable if there exists $F\colon X^n \to S$ satisfying $\mathcal{P}$ and generating $\mathfrak{A}$ according to the rule $\mathcal{R}$.

<u>Definition 1.5</u> Let $\mathcal{A}$ be a category and let $\mathfrak{p}$ be a property of objects in $\mathcal{A}$. We say <u>$\mathfrak{p}$ is intrinsic or $\mathcal{A}$ -intrinsic</u> if it is preserved by the equivalences of $\mathcal{A}$.

<u>Theorem 1.6</u> Let $\mathcal{A}$ be a real-generated category with datum $V = (\mathcal{A},\mathcal{P},\mathcal{R})$ and let $S \in \mathcal{S} \subset \mathcal{I}$. Then the property $[S,V,\mathcal{S}]$ - metrizable is intrinsic.

<u>Proof</u> We will omit the algebraic structure of objects. Let $(Y,\mathfrak{B})$ be $[S,V,\mathcal{S}]$ - metrizable and let $g\colon (X,\mathfrak{A}) \to (Y,\mathfrak{B})$ be an equivalence in $\mathcal{A}$. Then there exists a family $F\colon Y^n \to S$ satisfying $\mathcal{P}$ and $\mathfrak{B} = \mathfrak{B}_F$.

The family $G = \{f \circ g^n : f \in F\}$ satisfies $\mathcal{P}$ and $\mathfrak{U} = \mathfrak{U}_G$ because the properties $\mathcal{P}$ are natural for the category $\mathcal{R}$. Because $G: X^n \to S$ we see that $(X,\mathfrak{U})$ is $[S,V,\mathcal{S}]$ - metrizable. Therefore the property is intrinsic.

Let $\mathcal{H} = \{S \in \mathcal{S} : 0 \in S \subset [0,\infty)\}$. We may regard the sets in $\mathcal{H}$ as objects in a category. The morphisms, to be defined below, will play an essential role in the theorems of the later chapters.

Morphisms for objects $S \in \mathcal{H}$

Definition 1.7 If S and T are in $\mathcal{H}$ and $f: S \to T$ is a function we say f is in $\hom(S,T)$ if f satisfies

Hom 1 $f(0) = 0$,

Hom 2 $f(x) > 0$ if $x > 0$,

Hom 3 if x, y and z are in S and $x \leq y + z$ then $f(x) \leq f(y) + f(z)$, and

Hom 4 f is continuous at 0.

We obtain a category with objects S and morphisms $\hom(S,T)$.

If $d: X^2 \to S \subset R$ is a metric we say d is an S-metric on X.

<u>Theorem 1.8</u> If $f \in \hom(S,T)$ and d is an S-metric on the set X then $f \circ d$ is a metric on X which is uniformly equivalent to d, that is, $\mathfrak{U}_d = \mathfrak{U}_{f \circ d}$.

<u>Proof</u> The map $f \circ d$ is a metric by Hom 1, Hom 2 and Hom 3. Because f is continuous at 0, given $\varepsilon > 0$ there exists $\delta > 0$ such that if t is in S and $0 \leq t \leq \delta$ then $0 \leq f(t) \leq \varepsilon$. Thus if $d(x,y) \leq \delta$ then $0 \leq f(d(x,y)) \leq \varepsilon$. Conversely, if $\varepsilon > 0$ then $f(\varepsilon) > 0$ by Hom 2. Let $\delta = \frac{1}{2}f(\varepsilon)$. Then, if $0 \leq f(t) \leq \frac{1}{2}f(\varepsilon)$, $t < \varepsilon$ by Hom 3. Thus if $f(d(x,y)) \leq \delta$ then $d(x,y) \leq \varepsilon$. This completes the proof that d and $f \circ d$ are uniformly equivalent on X.

We might well have defined the morphisms as follows: if S and T are in $\mathcal{H}$ and $f: S \to T$ is a function and $\mathcal{Q}$ is a real generated category with datum $V = (\mathcal{Q}, \mathcal{P}, \mathcal{R})$, then $f \in \hom_{\mathcal{Q}}(S,T)$ if, for all $(X,\mathfrak{U}) \in \mathcal{Q}$ and $F: X^n \to S$ satisfying $\mathcal{P}$ and generating $\mathfrak{U}$ according to the rule $\mathcal{R}$, the family $f \circ F: X^n \to T$ satisfies $\mathcal{P}$ and generates $\mathfrak{U}$ according to the rule $\mathcal{R}$.

By Theorem 1.8, $\hom(S,T) \subset \hom_{\mathcal{U}}(S,T)$. The inclusion may be proper.

If $(X,\mathfrak{D},\mathfrak{U})$ is an object in a real-generated category we will frequently write X or $(X,\mathfrak{D})$ or $(X,\mathfrak{U})$ instead of $(X,\mathfrak{D},\mathfrak{U})$. The abbreviation used will be either specified or apparent from the context.

Theorem 1.9 Let $\mathcal{A}$ be a subcategory of the category of metrizable uniform spaces. Let $\mathcal{B}$ be a real-generated category which is derived from $\mathcal{A}$, let $\mathcal{H}$ be the family of subsets defined above and let $\mathcal{S} \subset \mathcal{H}$ be a non empty subfamily. Let $V_{\mathcal{A}}$ be the datum of $\mathcal{A}$ and $V_{\mathcal{B}}$ be the datum of $\mathcal{B}$. If S and T in $\mathcal{S}$ are such that $\hom(S,T) \neq 0$ then $[S,V_{\mathcal{B}},\mathcal{S}] < [T,V_{\mathcal{B}},\mathcal{S}]$.

Proof Let $(X,\mathfrak{U}) \in \mathcal{B}$. Because $\mathcal{B}$ is derived from $\mathcal{A}$ there exists a metric $d\colon X^2 \to R$ satisfying $\mathfrak{U} = \mathfrak{U}_d$. Let $(X,\mathfrak{U})$ be $[S,V_{\mathcal{B}},\mathcal{S}]$ - metrizable. Then there exists a metric d with range in S satisfying $\mathfrak{U} = \mathfrak{U}_d$. Let $f \in \hom(S,T)$. Then $f \circ d$ is a metric on X and $(X,\mathfrak{U}_d) = (X,\mathfrak{U}_{f\circ d})$ by Theorem 1.8. Therefore $(X,\mathfrak{U}_{f\circ d})$ is in $\mathcal{A}$ and, again because $\mathcal{B}$ is derived from

$\mathcal{A}$, $(X,\mathfrak{U}_d) = (X,\mathfrak{U}_{f\circ d})$. But $f \circ d: X^2 \to T \in S$. Therefore $(X,\mathfrak{U})$ is $[T,V_{\mathcal{B}},\mathcal{S}]$ - metrizable. This completes the proof.

This theorem is essentially a computational device. Later we will construct functions f in $\hom(S,T)$ for specific choices of S and T .

<u>Example 1.10</u> If S and T are in $\mathcal{S} \subset \mathcal{H}$ and satisfy $S \subset T$, and $\mathcal{A}$ is any real-generated category with datum V , then $[S,V,\mathcal{S}] < [T,V,\mathcal{S}]$.

<u>Example 1.11</u> If $a > 0$, V is the datum of a category satisfying the hypothesis of Theorem 1.9, and $S \in \mathcal{H}$, then

$$[a.S,V,\mathcal{H}] = [S,V,\mathcal{H}]$$

where $a.S = \{a.x : x \in S\}$.

<u>Example 1.12</u> $\hom(I,D) = \emptyset$ where $I = [0,1]$ and $D = \{0,1\}$ by Hom 1, Hom 2, and Hom 4.

We will examine the structure of the partially ordered sets $P(V,\mathcal{S})$, for various choices of category $\mathcal{A}$ with datum V and subfamilies $\mathcal{S} \subset \mathcal{I}$, in the chapters to follow.

<u>Example 1.13</u> Let $\mathcal{S} \subset \mathcal{I}$ have a largest member M when ordered by inclusion and let $\mathfrak{A}$ be any real-generated category with datum V . Then $[M,V,\mathcal{S}]$ is the maximum class of $P(V,\mathcal{S})$.

<u>Theorem 1.14</u> Let $S \in \mathcal{S}_1 \subset \mathcal{S}_2 \subset \mathcal{I}$. Then for each real-generated category $\mathfrak{A}$ with datum V ,

$$[S,V,\mathcal{S}_1] \subset [S,V,\mathcal{S}_2]$$

as sets of subsets of R .

<u>Theorem 1.15</u> Let $\mathcal{S} \subset \mathcal{I}$ be a non empty subfamily. Let $\mathfrak{A}$ with datum $V_{\mathfrak{A}}$ and $\mathfrak{B}$ with datum $V_{\mathfrak{B}}$ be real-generated categories such that $\mathfrak{B}$ is derived from $\mathfrak{A}$. Then for each S in $\mathcal{S}$,

$$[S,V_{\mathfrak{A}},\mathcal{S}] \subset [S,V_{\mathfrak{B}},\mathcal{S}]$$

as sets of subsets of R .

<u>Proof</u> Let S and T in $\mathcal{S}$ be such that $[S,V_{\mathfrak{A}},\mathcal{S}]$ is less than $[T,V_{\mathfrak{A}},\mathcal{S}]$. Let $(X,\mathfrak{B})$ in $\mathfrak{B}$ be such that there exists $F: X^n \to S$ satisfying $\mathfrak{B} = \mathfrak{B}_F$. Then necessarily $(X,\mathfrak{U}_F)$ is in $\mathfrak{A}$. There exists $G: X^n \to T$ satisfying $\mathfrak{U}_F = \mathfrak{U}_G$. Therefore $\mathfrak{B}_F = \mathfrak{B}_G$ because $\mathfrak{B}$ is

derived from $\mathcal{A}$. Thus $\mathfrak{B} = \mathfrak{B}_G$ and, because the functions G have values in T, $[S,V_{\mathcal{B}},\mathcal{S}] < [T,V_{\mathcal{B}},\mathcal{S}]$. Therefore if $T \in [S,V_{\mathcal{A}},\mathcal{S}]$ then $T \in [S,V_{\mathcal{B}},\mathcal{S}]$. This completes the proof.

Metric Spaces

Let $\mathcal{A}$ be a real-generated category with datum V . We may regard the rule $\mathcal{R}$ as a mapping from functions to structures

$$\mathcal{R}(F) = \mathfrak{U}_F$$

where $(X,\mathfrak{U}_F) \in \mathcal{A}$ and $F: X^n \to R$ satisfies $\mathcal{P}$. When $\mathcal{A}$ is $\mathcal{M}$, then category of metric spaces and isometries, the mapping is the "identity", $\mathcal{R}(d) = d$.

<u>Theorem 1.16</u> Let $\mathcal{P}$ be the usual axioms for metrics on sets. Let $V = (\mathcal{M},\mathcal{P},\mathcal{R})$ be the datum for the category of metric spaces and isometries. Then, for all S in $\mathcal{H}$,

$$[S,V,\mathcal{H}] = \{S\} .$$

<u>Proof</u> Let $T \in \mathcal{H}$ be such that $[T,V,\mathcal{H}] = [S,V,\mathcal{H}]$. Then $T < S$ and $S < T$. Because $S \in \mathcal{H}$, S is a

subset of $[0,\infty)$. Define a metric d on S by setting

$$d(x,y) = \begin{cases} \max\{x,y\} & \text{if } x \neq y \text{ , and} \\ 0 & \text{otherwise .} \end{cases}$$

Since $S \prec T$ and $d(S^2) \subset S$ there exists a metric g on S with $\mathfrak{M}_d = \mathfrak{M}_g$ and $g(S^2) \subset T$. But then necessarily $d = g$ and

$$S = d(S^2) = g(S^2) \subset T .$$

Similarly $T \subset S$. This completes the proof.

In general, if $S \in \mathcal{S} \subset \mathcal{H}$ then $[S,V,\mathcal{S}] = \{S\}$.

Hoop Structures

We will show that a metric d on a set X induces a structure on a subfamily of the power set of X which completely determines d . To this extent we may regard $(X,d) = (X,\mathfrak{M}_d)$ as being of the same theoretical type as the generated topological space $(X, \mathcal{T}_d)$ or generated uniform space $(X,\mathfrak{U}_d)$.

Although open balls in a metrizable space generate the topology it is not clear which properties of the open balls in a metric space characterize the metric form from which they were derived. Here a family of subsets called "hoops" is used to characterize the metric. What is more, when the hoops are derived from an S-metric, the hoops can be indexed by the elements of S .

Let (X,d) be a metric space and let

$$S(x,t) = \{y: d(x,y) = t\}$$

for each x in X and $t \geq 0$ in R . The family of subsets of X thus defined satisfies:

H1. For each $x \in X$, $X = \bigcup_{t \geq 0} S(x,t)$ where $S(x,0) = \{x\}$ and $S(x,t) \cap S(x,r) = \phi$ if $r \neq t$.

H2. If $x \in S(y,t)$ then $y \in S(x,t)$. Let $B(x,t) = \bigcup_{0 \leq r < t} S(x,r)$. Then

H3. $B(x,t) \supset \bigcup_{0 < r < t} \bigcup_{y \in S(x,r)} B(y,t-r)$.

<u>Definition 1.17</u> Conversely, we say a family $(h(x,r) : x \in X, r \geq 0)$ of subsets of a set X is a <u>hoop structure</u> $\mathfrak{H}$ <u>for</u> <u>X</u> if it satisfies H1, H2 and H3 with

$S(x,t)$ replaced by $h(x,t)$ wherever the former appears. It is clear that every metric space has a natural hoop structure which we will call the <u>hoop structure of the metric space</u>. When a set X is endowed with a hoop structure $\mathfrak{H}$ we will call it a hoop space $(X, \mathfrak{H})$.

Let $(X, \mathfrak{H})$ be a hoop space. We will show that the hoop structure $\mathfrak{H}$ is induced by a metric d on X. If x and y are in X let $d(x,y) = t$ where $x \in h(y,t)$. By H1, d is well defined; by H2, d is symmetric. If $d(x,y) = 0$ then $x = y$ and also the converse of this by H1 and H2. Also $h(x,y) = (y \in X : d(x,y) = t)$. It remains to prove that d satisfies the "triangle law."

Let x, y and z be distinct points in X with $d(x,z) = \ell$, $d(x,y) = m$ and $d(y,z) = n$. Let $\epsilon > 0$ be given and let $t = m + n + \epsilon$. Then $y \in h(x,m)$, $z \in h(y,n)$, $0 < m < n + m + \epsilon = t$ and $n < (m + n + \epsilon) - m = n + \epsilon$. Thus, by H3, $z \in B(x,t)$ and so $d(x,z) < t = m + n + \epsilon$. Therefore $d(x,z) \leq m + n = d(x,y) + d(y,z)$. We have completed the proof of

<u>Theorem 1.18</u> The hoop structure of a metric space determines the metric uniquely and vice versa.

This discussion applies when hoops are labelled by elements of R^+. If an S-metric space is being considered, the hoops can be indexed by elements of S. It serves our purposes, however, to use arbitrary positive real numbers as the radii of balls. The hoop axioms then become:

<u>Definition 1.19</u> An <u>S-hoop</u> <u>structure</u> on a set X is a family $(h(x,r) : x \in X , r \in S)$ of subsets of X satisfying:

S1. $X = \bigcup_{t \in S} h(x,t)$ where $h(x,0) = \{x\}$ and

$h(x,t) \cap h(x,r) = \phi$ if $t \neq r$,

S2. If $x \in h(y,t)$ then $y \in h(x,t)$. Define

$B(x,t) = \bigcup \{h(x,r) : r \in S , 0 \leq r < t\}$ when

$t \in (0,\infty)$. Then

S3. $B(x,t) \supset \bigcup_{\substack{r \in S \\ 0 < r < t}} \bigcup_{y \in h(x,r)} B(y,t-r)$.

<u>Theorem 1.20</u> The S-hoop structure of an S-metric space determines the metric uniquely and vice versa.

Metrizable convergence and proximity spaces

Let $\mathcal{P}$ be the usual metric axioms and let $\mathcal{S} \subset \mathcal{H}$ be a non empty subfamily. Let $\mathcal{T}$ be the category of metrizable topological spaces and continuous functions. Let the rule $\mathcal{R}_{\mathcal{T}}$ be the usual rule for the generation of a topology by a metric. Let $\mathcal{C}$ be the category of metrizable convergence spaces and continuous maps. The rule $\mathcal{R}_{\mathcal{C}}$ for the generation of convergence structures is as follows: if X is a set, $d: X^2 \to R$ a metric satisfying $\mathcal{P}$, x an element of X , and $\mathcal{F}$ a filter on X , then $\mathcal{F}$ converges to x if and only if, for all $\epsilon > 0$, the ball $B(x,\epsilon) \in \mathcal{F}$. Let $V_{\mathcal{T}} = (\mathcal{T}, \mathcal{P}, \mathcal{R}_{\mathcal{T}})$ be the datum of $\mathcal{T}$ and $V_{\mathcal{C}} = (\mathcal{C}, \mathcal{P}, \mathcal{R}_{\mathcal{C}})$ be the datum of $\mathcal{C}$. Then

<u>Theorem 1.21</u> For all $S \in \mathcal{S}$, $[S, V_{\mathcal{T}}, \mathcal{S}] = [S, V_{\mathcal{C}}, \mathcal{S}]$.

<u>Proof</u> By Theorem 1.15, because $\mathcal{C}$ is derived from $\mathcal{T}$, we have $[S, V_{\mathcal{T}}, \mathcal{S}] \subset [S, V_{\mathcal{C}}, \mathcal{S}]$. Let $T \in [S, V_{\mathcal{C}}, \mathcal{S}]$. Then, by an abuse of language, $S < T$ and $T < S$ in $\mathcal{C}$. We will prove that $S < T$ and $T < S$ in $\mathcal{T}$: let $(X, \mathcal{J})$ in $\mathcal{T}$ be such that there exists a metric d on X with $d(X^2) \subset T$ and $\mathcal{J} = \mathcal{J}_d$. Then $(X, \mathfrak{C}_{\mathcal{J}})$, the convergence space derived from the topological space $(X, \mathcal{J})$, is equal

to $(X,\mathfrak{C}_d)$. Because $T < S$ in $\mathcal{B}$ there exists a metric g on X such that $g(X^2) \subset S$ and $\mathfrak{C}_g = \mathfrak{C}_d$. But then $\mathfrak{T}_g = \mathfrak{T}_d$ because $\mathcal{B}$ can be regarded as a full subcategory of $\mathcal{J}$. Similarly, because $S < T$ in $\mathcal{B}$, $S < T$ in $\mathcal{J}$. Therefore $T \in [S,V_{\mathcal{J}},\mathcal{S}]$ and so, for all S , $[S,V_{\mathcal{J}},\mathcal{S}] = [S,V_{\mathcal{B}},\mathcal{S}]$.

Note that, by the above theorem, $P(V_{\mathcal{J}},\mathcal{S}) = P(V_{\mathcal{B}},\mathcal{S})$.

If $d: X^2 \to R$ is a metric on a set X and A and B are subsets of X , we say $A \delta B$ (A is near to B) if $d(A,B) = 0$. This rule, denoted $\mathcal{R}_{\mathcal{N}}$, defines a proximity structure, $\mathfrak{N}_d$, on the set X . Then the datum V for the category of metrizable proximity spaces $\mathcal{N}$ and p-continuous maps is given by $V_{\mathcal{N}} = (\mathcal{N},\mathcal{P},\mathcal{R}_{\mathcal{N}})$. Let $V_{\mathcal{U}}$ be the usual datum for the category $\mathcal{U}$ of metrizable uniform spaces. Then

<u>Theorem 1.22</u> $[S,V_{\mathcal{U}},\mathcal{S}] = [S,V_{\mathcal{N}},\mathcal{S}]$ for all $S \in \mathcal{S} \subset \mathcal{H}$.

<u>Proof</u> The proof is similar to that of the previous theorem because $\mathcal{N}$ is derived from $\mathcal{U}$, and, if $\mathfrak{N}_d = \mathfrak{N}_g$, then $\mathfrak{U}_d = \mathfrak{U}_g$. This in turn follows from the following

theorem:

"Let $(X,\mathfrak{N})$ be a proximity space. Then the family of uniformities on X which generate $\mathfrak{N}$ contains at most one pseudometrizable uniformity", Thron [65, Theorem 21.28, page 203].

Therefore if d and g generate $\mathfrak{N}$ they must induce the same uniform structure on X.

Invariants on a Category

Definition 1.23 Let $\mathfrak{A}$ and $\mathfrak{B}$ be categories and let I be a mapping from the objects of $\mathfrak{A}$ to the objects of $\mathfrak{B}$. We say that I is an invariant on $\mathfrak{A}$ if $A \cong B$ in $\mathfrak{A}$ implies $I(A) \cong I(B)$ in $\mathfrak{B}$. We say that I is a proper invariant on $\mathfrak{A}$ if there exist A and B in $\mathfrak{A}$ such that $I(A) \not\cong I(B)$. Otherwise we say that I is an improper invariant. We say that I is a faithful invariant if, for any pair of objects A and B in $\mathfrak{A}$, if $A \not\cong B$ then $I(A) \not\cong I(B)$.

Any functor is an invariant. The identity functor is a faithful invariant.

<u>Example 1.24</u> Let $\mathcal{A}$ be the category of compact surfaces and continuous functions. Then the fundamental group is a faithful invariant, $\pi : \mathcal{A} \to \mathcal{B}$, where $\mathcal{B}$ is the category of groups and homomorphisms.

<u>Example 1.25</u> The homology functor is an improper invariant on the category of countable metric spaces.

Let $\mathcal{A}$ be a real-generated category with datum V , let $\mathcal{S} \subset \mathcal{J}$ and let $(X,\mathfrak{U}) \in \mathcal{G}$. We can consider the class $[S,V,\mathcal{S}]$ to be an invariant I from $\mathcal{A}$ to a category $\mathcal{P}$ defined as follows: There are two objects $\{T,F\}$ and the only morphisms are the identities. Then let $I(X) = T$ if $(X,\mathfrak{U})$ is $[S,V,\mathcal{S}]$ - metrizable and let $I(X) = F$ otherwise.

<u>Example 1.26</u> If $\mathcal{S} = \{\{0\}\}$ then $[S,V,\mathcal{S}]$ is an improper invariant for each category with datum V .

<u>Example 1.27</u> Let $\mathcal{A}$ be the category of metrizable topological spaces which are connected and have more than one point. Let $\mathcal{S} \subset \mathcal{H}$ and let $V = (\mathcal{A}, \mathcal{P}, \mathcal{R})$, where $\mathcal{R}$ is the usual rule for the generation of a metric space topology. Then $[S,V,\mathcal{S}]$ is an improper invariant for all $S \in \mathcal{S}$.

In the sequel we will examine those derived categories for which the classes correspond to proper invariants.

The construction of morphisms

In this section we will construct functions in $\hom(S,T)$ for specific choices of S and T in $\mathcal{H}$.

Let $\{a_n\}$ and $\{b_n\}$ be strictly monotonically decreasing sequences of real numbers with limit zero. Let $\lambda = (\{a_n\}, \{b_n\})$. Then λ determines the graph of a piecewise linear continuous function f_λ as follows: in Euclidean 2-space join (a_n, b_n) to (a_{n+1}, b_{n+1}) with a straight line for $n = 1, 2, \ldots$. Finally, let $f_\lambda(0) = 0$. Then $f_\lambda: [0, a_1] \to [0, b_1]$ has the property that $x < y$ if and only if $f_\lambda(x) < f_\lambda(y)$. We will say that the pair of sequences λ determines the function f_λ and let L denote the class of all such functions. Let L_S be the set of functions in L which are subadditive.

Theorem 1.28 Let f in L be determined by the pair of sequences $(\{a_n\}, \{b_n\})$. If $\dfrac{b_n}{a_n} \leq \dfrac{b_{n+1}}{a_{n+1}}$ for all n in $\mathbb{N}$ (the natural numbers), then $f \in L_S$.

Proof Let f be in L and suppose that

$\frac{b_n}{a_n} \leq \frac{b_{n+1}}{a_{n+1}}$ for all n in $\mathbb{N}$. We will prove that

$\frac{f(x)}{x}$ is decreasing on $(0, a_1]$. This will be sufficient to show that f is subadditive [28, Theorem 7.2.4, page 239]. Consider the restriction of f to $[a_{n+1}, a_n]$. the graph of f has the equation $f(x) = m_n x + c_n$ where

$$m_n = \frac{b_n - b_{n+1}}{a_n - a_{n+1}} \quad \text{and} \quad c_n = \frac{a_n b_{n+1} - a_{n+1} b_n}{a_n - a_{n+1}} .$$

Let $\{x_1, x_2\} \subset [a_{n+1}, a_n]$. Then, if $x_1 \leq x_2$,

$\frac{f(x_2)}{x_2} \leq \frac{f(x_1)}{x_1}$ if and only if $m_n + \frac{c_n}{x_2} \leq m_n + \frac{c_n}{x_1}$,

if and only if $c_n \geq 0$, which is true if and only if

$\frac{b_n}{a_n} \leq \frac{b_{n+1}}{a_{n+1}}$. Therefore, if $\frac{b_n}{a_n} \leq \frac{b_{n+1}}{a_{n+1}}$ then $\frac{f(x)}{x}$ is decreasing on $(0, a_1]$ because f is continuous. Thus f is subadditive.

Theorem 1.29 If f in L is determined by the pair $(\{a_n\}, \{b_n\})$ then there is a subsequence $\{a_{n_j}\}$ of $\{a_n\}$ such that the element of L determined by $(\{a_{n_j}\}, \{b_j\})$ is in L_S.

Proof Let $\lambda = (\{a_n\}, \{b_n\})$. We will define a sequence of piecewise linear functions $\{f_n\}$ as follows: let f_1 be the linear function joining $(0, 0)$ and (a_1, b_1) and let $n_1 = 1$. Because $a_n \to 0$ there exists an $n_2 > n_1$ such that $\frac{b_2}{a_{n_2}} \geq \frac{b_1}{a_{n_1}}$. Join $(0, 0)$ to (a_{n_2}, b_2) and (a_{n_2}, b_2) to (a_{n_1}, b_1) with straight lines thus defining a function f_2 . By induction choose $n_{j+1} > n_j$ such that $\frac{b_{j+1}}{a_{n_{j+1}}} \geq \frac{b_j}{a_{n_j}}$, and form the piecewise linear function f_n . By Theorem 1.28 each f_n is subadditive on $[0, a_1]$ and thus $g(x) = \sup_n \{f_n(x)\}$ is subadditive. Furthermore g is determined by $(\{a_{n_j}\}, \{b_j\})$ and is in L_S .

The following result is of fundamental importance for this theory.

Theorem 1.30 Let $S \in \mathcal{H}$ be countable and dense in $[0, 1]$, let $T \in \mathcal{H}$ be dense in $[0, 1]$, let $1 \in S \cap T$, and let $S \cup T \subset [0, 1]$. Then $\hom(S,T) \neq \phi$.

Proof We will define an increasing sequence (f_n) of piecewise linear (PL) functions in $\hom([0,1],[0,1])$ and prove that $f = \sup f_n$, when restricted to S , is in $\hom(S,T)$. The points of S will be labelled, as the proof proceeds, in two ways as follows:

Points of Type I: If $S \setminus \{0\} = \{s_i : 0 \leq i < \infty\}$, $(s_o = 1)$, points of type I will be a subsequence (s_{n_i}) where $n_o = 0$, $n_{i+1} = n_i + i + 1$ for $i \geq 0$,

$$s_{n_o} > s_{n_1} > s_{n_2} > \ldots > 0 \text{, and } s_{n_i} \to 0 .$$

Points of Type II: The points in $S \cap (s_{n_i}, s_{n_{i-1}})$ are labelled $\{t_{ij} : i \leq j < \infty\}$. Then we let

$$s_{n_i+1} = t_{1i}, \; s_{n_i+2} = t_{2i}, \; \ldots, \; s_{n_i+i} = t_{ii} .$$

Keeping this scheme is mind we will define the sequence (f_n):

(a) Definition of f_o: This is the PL function whose graph is the straight line joining (0, 0) to (1, 1). Clearly $f_o \in \hom([0, 1], [0, 1])$.

(b) Definition of f_{n_1}: The set

$$[0,\tfrac{1}{2}] \times [0,\tfrac{1}{2}] \cap \{(x,y) : y > f_o(x)\} \cap S \times T \neq \phi .$$

Choose a point in this set and label it (s_{n_1}, r_{n_1}). Let f_{n_1} be the PL function consisting of straight lines joining (0, 0), (s_{n_1}, r_{n_1}) and (1, 1). Then f_{n_1} is in

Figure 1

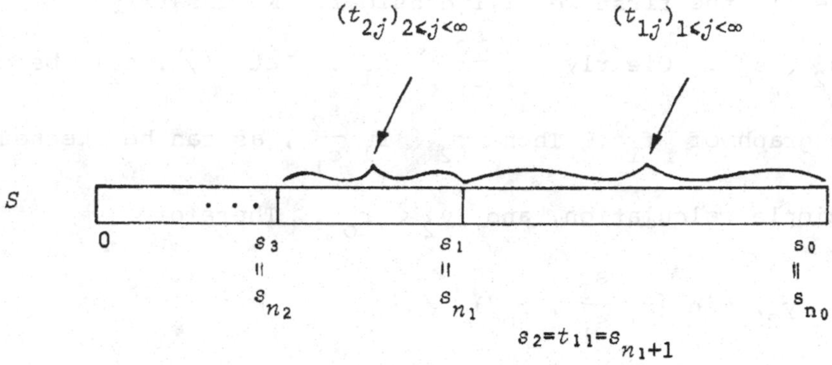

Labelling the points of S

hom([0, 1], [0, 1]) .

We then label all the points in $(s_{n_1}, 1) \cap S$ by $\{t_{1j}: 1 \leq j < \infty\}$.

(c) Definition of $f_2 = f_{n_1+1}$: Now $s_2 = t_{11}$ is a point of type II, the first we will consider. Necessarily $s_1 < s_2 < s_o$. Clearly $r_1 \frac{s_2}{s_1} > r_1$. Let (y_2, s_2) be on the graph of f_1 . Then $y_2 < r_1 \frac{s_2}{s_1}$, as can be checked by a simple calculation, and $y_2 < r_o$. Therefore

$$(y_2, \min \{r_1 \frac{s_2}{s_1} , r_o\}) \neq \phi$$

and we choose r_2 to be a point in T satisfying $y_2 < r_2 < \min \{r_1 \frac{s_2}{s_1} , r_o\}$. Then $r_1 < r_2 < r_o$ and $\frac{r_1}{s_1} > \frac{r_2}{s_2} > \frac{r_o}{s_o}$. Then f_{n_2} is the PL subadditive function which joins $(0, 0)$, (s_1, r_2) , (s_2, r_2) and (s_o, r_o) with straight line sections.

(d) Definition of $f_{n_{i+1}}$: Suppose $f_o, \ldots, f_{n_i+i}$ have been defined. Then

$$S \times T \cap \{(x,y): y > f_{n_i}(x)\} \cap (0,2^{-i}) \times (0,2^{-i}) \cap$$

$$(0, s_{n_i}) \times (0, r_{n_i}) \neq \phi .$$

A point is chosen in this set and labelled $(s_{n_{i+1}}, r_{n_{i+1}})$.

Figure 2

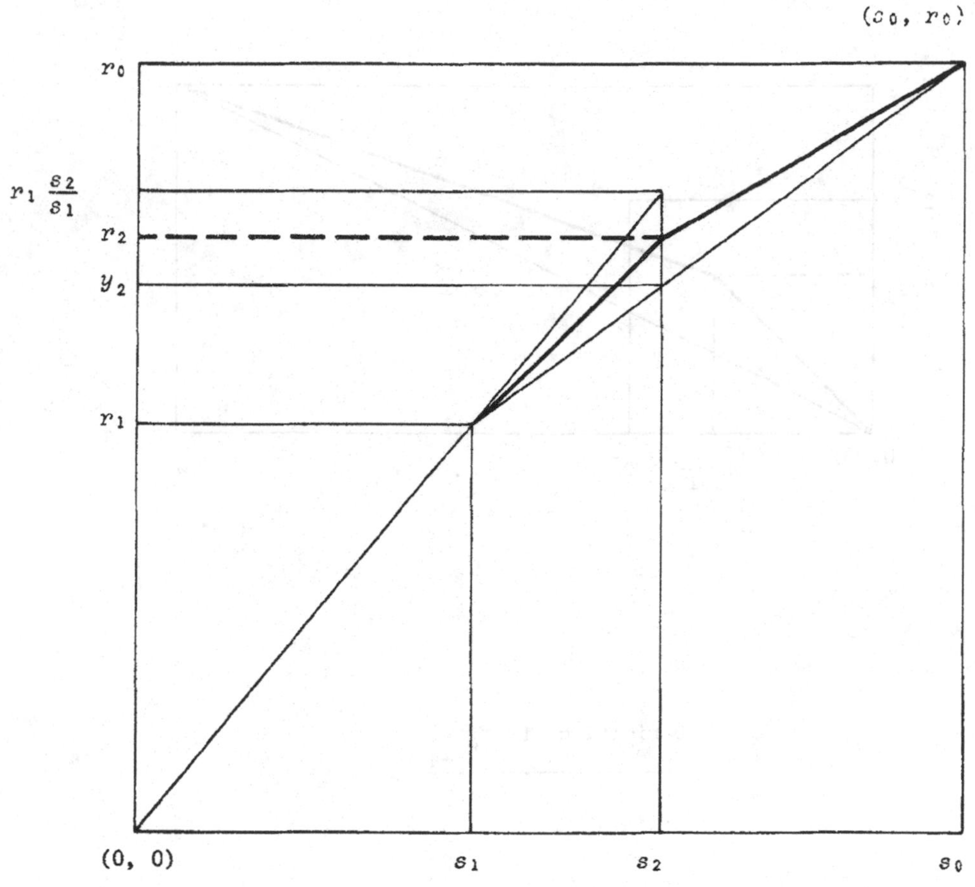

Definition of f_2

Figure 3

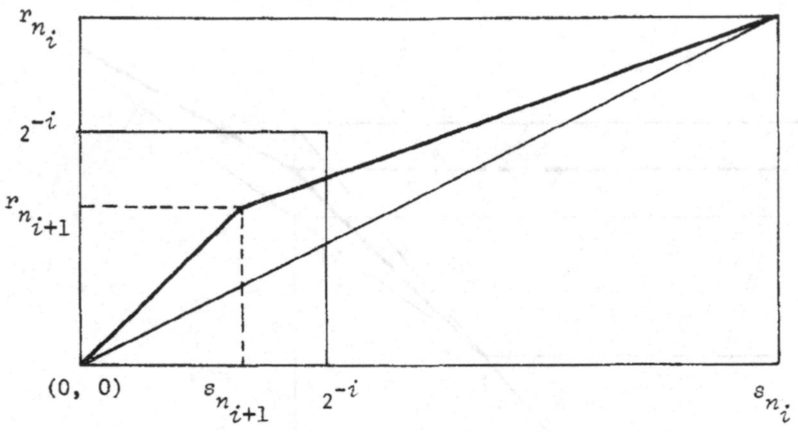

Definition of $f_{n_{i+1}}$

(See figure 3.) When $s_{n_i} \leq x \leq s_o$, let $f_{n_{i+1}}(x) = f_{n_i+1}(x)$. When $0 \leq x \leq s_{n_i}$ let $f_{n_{i+1}}$ be the PL function joining (0, 0), $(s_{n_{i+1}}, r_{n_{i+1}})$ and (s_{n_i}, r_{n_i}). Then $f_{n_{i+1}}$ is in hom([0, 1], [0, 1]).

Label the points in $(s_{n_{i+1}}, s_{n_i}) \cap S$ by $\{t_{i+1,j}: i+1 \leq j < \infty\}$ and we now proceed to define f_k when s_k is a point of type II according to the labelling scheme outlined above.

(e) Definition of f_k when s_k is a point of type II: There exists n and j in $\mathbb{N}$ with $n, j < k$ and $s_i < s_k < s_j$ and $(s_i, s_j) \cap \{s_\ell: \ell < k\} \neq \phi$. Define $f_k(x) = f_{k-1}(x)$ when $0 \leq x \leq s_i$ or $s_j \leq x \leq s_o$, and as the PL function joining (s_i, r_i), (s_k, r_j) and (s_j, r_j) otherwise, where r_k is chosen in a similar way to the way we chose r_2 in (c) above. (See figure 4.)

This completes the inductive definition of the sequence (f_n).

The sequence has the following properties:

(i) each f_n is strictly increasing,

(ii) $f_n(s_i) = r_i$ for $n \geq i$,

Figure 4

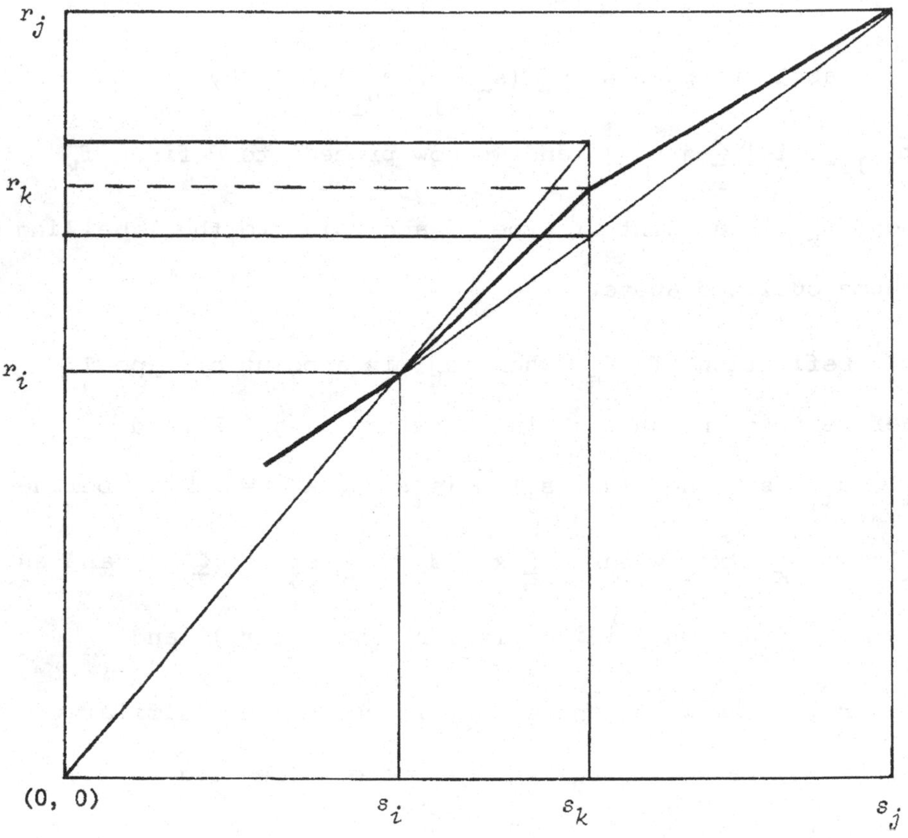

Definition of f_k

(iii) $f_n \leq f_{n+1}$ for all n ,

(iv) $f_n(0) = 0$ for all n , and

(v) $s_{n_{i+1}} < 2^{-i}$ and $r_{n_{i+1}} < 2^{-i}$.

Let $f(x) = \sup_n f_n(x)$ for each x in $[0, 1]$. Then f is subadditive, $f(x) > 0$ if $x > 0$, $f(0) = 0$, $f(s_i) = r_i$ for $0 \leq i < \infty$, f is increasing, and finally, f is continuous at zero (because $s_{n_i} \to 0$, $f(s_{n_i}) \to 0$ and f is increasing). Therefore f restricted to S is in $\hom(S,T)$. This completes the proof.

<u>Theorem 1.31</u> Let (a_n) , (b_n) be a pair of sequences with $a_1 \geq b_1 > a_2 \geq b_2 > \ldots > 0$, $a_n \to 0$, and

$$\frac{a_2 \ldots a_n}{b_1 \ldots b_{n-1}} = c_n \to 0 \text{ where } c_1 = 1 .$$

Let $S = \bigcup_{n=1}^{\infty} [b_n, a_n] \cup \{0\}$ and $T = \{c_n : n \in \mathbb{N}\} \cup \{0\}$. Then there exists an f in $\hom([0,a_1] , [0, 1])$ with $f([b_n, a_n]) = \{c_n\}$ for each n in $\mathbb{N}$.

<u>Proof</u> Define f inductively on subintervals of $[0, a_1]$ as follows: if $b_1 \leq x \leq a_1$ let $f(x) = c_1 = 1$. If $a_2 \leq x \leq b_1$ the graph of f is the straight line

Figure 5

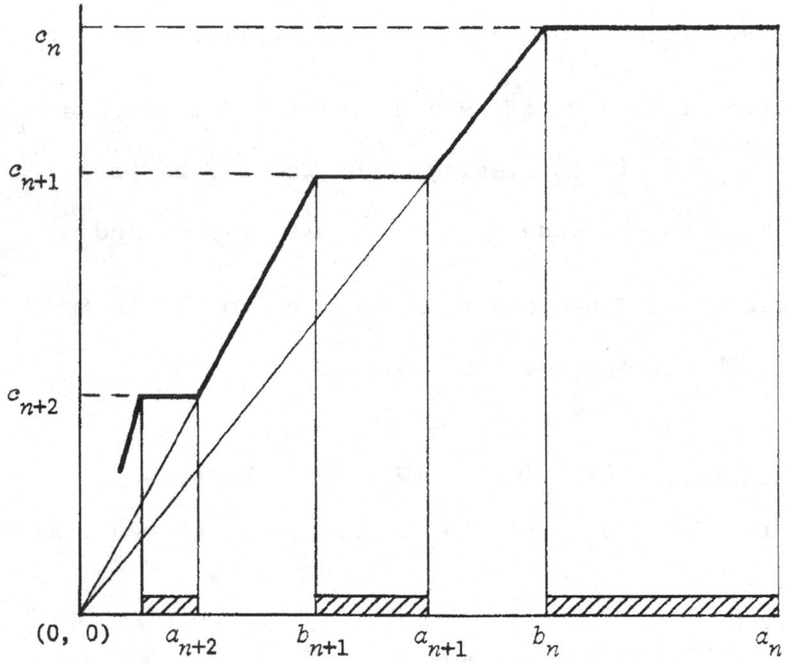

Subadditive Steps

segment joining (a_2, c_2) to (b_1, c_1). If $b_2 \leq x \leq a_2$ let $f(x) = c_2$. After defining f on $[b_n, a_n]$ to have the constant value c_n we define f on $[a_{n+1}, b_n]$ to be the straight line joining (a_{n+1}, c_{n+1}) to (b_n, c_n). (See figure 5.) Then

$$\frac{c_{n+1}}{a_{n+1}} = \frac{c_n}{b_n} \geq \frac{c_n}{a_n} \quad \text{for each } n \text{ in } \mathbb{N} .$$

Therefore, by Theorem 1.28, f is subadditive on $(0, a_1]$. Finally let $f(0) = 0$. Because $c_n \to 0$, $f \in \hom([0, a_1], [0, 1])$. Clearly, $f([b_n, a_n]) = \{c_n\}$ completing the proof.

<u>Theorem 1.32</u> If $f \in \hom(I, I)$ is strictly increasing and $T, S \subset (0, 1]$ are such that $T \subset f(S)$ then $f \in \hom(I \setminus S, I \setminus T)$.

<u>Proof</u> We need only remark that if x is not in S and f is injective then $f(x)$ is not in $f(S)$, which means $f(x)$ is not in T.

<u>Theorem 1.33</u> Let $T \subset (0, 1]$ be countable and let $S \subset (0, 1]$ be dense. Then there exists an f in $\hom(I, I)$ which is strictly increasing and such that $T \subset f(S)$.

Proof The proof of this theorem is very similar to the proof of Theorem 1.30 and it will not be given here. We need only remark that the points of T are indexed first rather than the points of S. Theorem 1.30 gives a strictly increasing function and the same is true in this case.

Theorem 1.34 Let S and T be in $\mathcal{H}$. If zero is a cluster point of S and $f \in \hom(S,T)$ then f is uniformly continuous on S.

Proof Let $\varepsilon > 0$ be given and let (s_n) be a sequence of distinct points in S with $s_n \downarrow 0$. Because f is continuous at 0 there is an n in $\mathbb{N}$ with $f(s_n) < \varepsilon$. Then, if x and y in S satisfy $0 \leq x - y \leq s_n$, it follows that $0 \leq f(x) - f(y) \leq f(s_n) < \varepsilon$. Thus f is uniformly continuous.

Let $\mathbb{Q}$ be the rational numbers and $\mathbb{Q}^+ = \mathbb{Q} \cap [0,\infty)$. Let $\mathbb{P}$ be the irrational numbers and $\mathbb{P}^+ = \{\mathbb{P} \cap [0,\infty)\} \cup \{0\}$

Theorem 1.35 $\hom(\mathbb{P}^+, \mathbb{Q}^+) = \phi$.

Proof If $f \in \hom(\mathbb{P}^+, \mathbb{Q}^+)$ then $f: \mathbb{P}^+ \to \mathbb{Q}^+$ is uniformly continuous. Therefore f has a unique continuous extension to a function $g: \mathbb{R}^+ \to \mathbb{R}^+$. Then

$$g(\mathbb{R}^+) \subset g(\mathbb{P}^+) \cup g(\mathbb{Q}^+)$$

which is countable. Connectedness then implies $f(\mathbb{P}^+) = \{0\}$, which is a contradiction.

Let $\mathbb{N} = \{1, 2, 3, \ldots\}$, $H = \{1/n: n \in \mathbb{N}\} \cup \{0\}$, $D_m = D_{m-1} + H$ where $D_1 = H$ and $m \geq 2$, and $J_m = \{I - D_m\} \cup \{0\}$ for $m \in \mathbb{N}$ and $I = [0, 1]$.

<u>Theorem 1.36</u> For each m in $\mathbb{N}$, $\hom(J_m, J_{m+1}) = \phi$.

<u>Proof</u> Let $f \in \hom(J_m, J_{m+1})$ and let g be the extension of f to a function in $\hom(I, I)$. Because g is continuous we might as well assume that the range of g is all of I . Let $x \in D_{m+1}$. Then necessarily $g^{-1}(x) \subset D_m$ as the range of f is in J_{m+1} . If a and b are in $g^{-1}(x)$ then $a = b$. Clearly $x < y$ in D_{m+1} if and only if $g^{-1}(x) < g^{-1}(y)$ in D_m . Therefore the assignment $x \to g^{-1}(x)$ is a homeomorphism onto its range in D_m . But this is impossible as can be seen by considering the iterated derived sets of D_{m+1} and D_m . This contradiction completes the proof.

Theorem 1.37 For each m in $\mathbb{N}$, $\hom(J_m, \mathbb{P}^+) = \phi$

Proof As in the preceding theorem extend f to a function g in $\hom(I, I)$ and suppose that the range of g is I. Then apply a very similar argument to the set $g^{-1}(D_{m+1}) \subset D_m$.

Theorem 1.38 If $S \in \mathcal{H}$ is dense in a neighborhood of zero and $T \in \mathcal{H}$ is not dense in any neighborhood of zero, then $\hom(S, T) = \phi$.

Proof Let S be dense in $[0, \delta]$ and let $f \in \hom(S \cap [0, \delta], T)$. If g is the continuous extension of f to a function in $\hom([0, \delta], \overline{T})$ then $g([0, \delta]) = \{0\}$ necessarily. This contradiction completes the proof.

Let $H = \{1/n : n \in \mathbb{N}\} \cup \{0\}$. Then

Corollary 1.39 $\hom(\mathbb{Q}^+, H) = \phi$.

Chapter II : Uniform Spaces

We have shown that when $\mathcal{a}$ is a real-generated category with datum V and $\mathcal{S} \subset \mathcal{J}$ then $P(V, \mathcal{S})$ is a partially ordered set. In this chapter we will consider these partially ordered sets in detail for several subcategories $\mathcal{a}$ of the category $\mathcal{U}$ of metrizable uniform spaces and for the category $\mathcal{E}$ of all uniform spaces.

S - metrizable uniform spaces

In the first part of this chapter we have adopted the following conventions: families of subsets $\mathcal{S}$ of R are non empty subfamilies of $\mathcal{H}$. The rule $\mathcal{R}$ for the generation of uniform structures by metrics satisfying $\mathcal{P}$ is the usual rule. If $V = (\mathcal{a}, \mathcal{P}, \mathcal{R})$ is the datum of $\mathcal{a}$ we frequently write $P(\mathcal{a}, \mathcal{S})$ instead of $P(V, \mathcal{S})$ and $[S, \mathcal{a}, \mathcal{S}]$ instead of $[S, V, \mathcal{S}]$ when $S \in \mathcal{S}$. The expression S-metrizable means $[S, V, \mathcal{S}]$-metrizable.

Let $D = \{0,1\}$. Then

Theorem 2.1 Let $S \in \mathcal{H}$ be such that 0 is not a cluster point of S . If $(X, \mathcal{U})$ is S-metrizable then

$\Delta \in \mathfrak{U}$. Conversely, if $(X,\mathfrak{U})$ is metrizable and $\Delta \in \mathfrak{U}$, then $(X,\mathfrak{U})$ is D-metrizable.

Proof Let d be a metric on X satisfying $\mathfrak{U}_d = \mathfrak{U}$ and $d(X^2) \subset S$. Then there exists a $\delta > 0$ such that if $d(x,y) < \delta$ then $x = y$. Thus $\Delta = \{(x,y) : d(x,y) < \delta\} \in \mathfrak{U}$. In proving the converse note that we do not need to assume that $(X,\mathfrak{U})$ is metrizable.

Let $W = \{0\} \cup \{3^{-n}: n \in \mathbb{N}\}$.

Theorem 2.2 The uniform space $(X,\mathfrak{U})$ is W-metrizable if and only if there exists a base $(U_n)_{n \in \mathbb{N}}$ for $\mathfrak{U}$ satisfying $U_n^2 = U_n$ for all n in $\mathbb{N}$.

Proof (⇒) Let d be a W-metric on X satisfying $\mathfrak{U}_d = \mathfrak{U}$ and let $U_n = \{(x,y) : d(x,y) < 3^{-n}\}$ for each n in $\mathbb{N}$. Then (U_n) is a base for $\mathfrak{U}$ and $U_n^2 = U_n$ for all n .

(⇐) If (U_n) has the given property let $h(x,y) = 0$ if $x = y$ and $h(x,y) = \min \{3^{-n}: (x,y) \in U_n\}$ if $x \neq y$. Then h is a W-metric on X and $\mathfrak{U}_h = \mathfrak{U}$.

The conclusion of the following Theorem is weaker than that of Theorem 2.4 below. We have included it because the method of proof is different.

<u>Theorem 2.3</u> Let S be a closed subset of $[0,\infty)$ with $0 \in S$ and $\mathrm{Ind}\, S = 0$. Let (X,d) be an S-metric space. Then there is an H-metric on X which generates the same uniform structure as d.

<u>Proof</u> Let $\gamma > 0$ be in S and $d'(x,y) = \min\{d(x,y), \gamma\}$ for each x, y in X. Then d' is an S-metric on X generating the same uniform structure as d. We may therefore assume that S is a compact subset of $[0,\infty)$ and drop the dash from d'.

Let $\rho: S \times S \to H$ be an H-metric on S compatible with the topology S inherits as a subspace of $[0,\infty)$, Broughan [13]. The metric ρ is uniformly continuous on $S \times S$. Define a metric $\theta: X \times X \to H$ by the formula

$$\theta(x,y) = \max_{x_0 \in X} \rho(d(x, x_0), d(x_0, y)) .$$

We will show that θ generates the same uniform structure as d.

(a) The identity, $\mathrm{id}: (X,d) \to (X,\theta)$ is uniformly continuous: Given $\epsilon > 0$ there is a $\delta > 0$ such that if

ℓ and m are in S and $|\ell - m| < \delta$ then $\rho(\ell,m) < \varepsilon/2$. When x, y, x_o are in X , $|d(x, x_o) - d(x_o, y)| \leq d(x,y)$. Therefore, if $d(x,y) < \delta$ then $|d(x, x_o) - d(x_o, y)| < \delta$ which implies $\rho(d(x, x_o), d(x_o, y)) < \varepsilon/2$ for all x_o in X . Thus $\max_{x_o \in X} \rho(d(x, x_o), d(x_o, y)) < \varepsilon$ and hence id: $(X,d) \to (X,\theta)$ is uniformly continuous.

(b) Conversely, id: $(X,\theta) \to (X,d)$ is uniformly continuous: Suppose there is an $\varepsilon_o > 0$ and a sequence of pairs $(x_n, y_n) \in X^2$ such that, for all n in $\mathbb{N}$, $\theta(x_n, y_n) < \frac{1}{n}$ and $d(x_n, y_n) \geq \varepsilon_o$. Then for all x_o in X and n in $\mathbb{N}$,

$$\rho(d(x_n, x_o), d(y_n, x_o)) < \frac{1}{n} .$$

Choose $x_o = x_n$. Then $\rho(0, d(x_n, y_n)) < \frac{1}{n}$. If we let $d(x_n, y_n) = a_n$ then $\{a_n\} \subseteq S \cap [\varepsilon_o, \infty)$ which is compact. There is a subsequence $a_{n_j} \to a \geq \varepsilon_o$. Then $\rho(0, a_{n_j}) \to \rho(0,a) \neq 0$. But $\rho(0, a_{n_j}) < \frac{1}{n_j}$ for all j in $\mathbb{N}$. This contradiction shows that id: $(X,\theta) \to (X,d)$ is uniformly continuous. Then from (a) and (b), θ and d generate the same uniform structure on X .

<u>Theorem 2.4</u> Let $S \in \mathcal{H}$ be such that cl(S) (the closure of S in $\mathbb{R}^+$) is not a neighborhood of 0 in $\mathbb{R}^+$. If $(X,\mathfrak{U})$ is S-metrizable then $(X,\mathfrak{U})$ is H-metrizable.

<u>Proof</u> Because $[H,\mathcal{U},\mathcal{H}] > [W,\mathcal{U},\mathcal{H}]$ (example 1.10) and Theorem 2.2 we need only find a base (W_n) for $\mathfrak{U}$ satisfying $W_n^2 = W_n$ for all n in $\mathbb{N}$.

Because cl(S) is not a neighborhood of 0 there exist monotonic sequences (a_n) and (b_n) in $\mathbb{R}^+$ satisfying $[a_n, b_n] \cap S = \emptyset$ for all n, $0 < a_n < b_n$ for all n, and $\lim_{n\to\infty} a_n = \lim_{n\to\infty} b_n = \lim_{n\to\infty} (b_n - a_n) = 0$.

Let $U_n = \{(x,y) : d(x,y) \leq a_n\}$ where d is an S-metric on X generating the uniform structure $\mathfrak{U}$. Let $n_1 = 1$. There exists an $n_2 > n_1$ such that $a_{n_2} < b_{n_1} - a_{n_1}$. Then $U_{n_2} \circ U_{n_1} = U_{n_2} \circ U_{n_2} = U_{n_1}$. Proceeding inductively, there exists an $n_j > n_{j-1}$ such that $U_{n_j} \circ U_{n_{j-1}} = U_{n_{j-1}} \circ U_{n_j} = U_{n_{j-1}}$ for all $j \geq 2$.

Let $K_j = U_{n_j}$. Then $U = (U_n)_{n\in\mathbb{N}} = (K_j)_{j\in\mathbb{N}}$, that is to say, (K_j) is a base for $\mathfrak{U}$. If $j \in \mathbb{N}$ then $K_{j+1}^2 \subset K_{j+1} \circ K_j = K_j$. By induction $K_{j+1}^n \subset K_j$ for each n in $\mathbb{N}$ and therefore

$$\bigcup_{n=1}^{\infty} K_{j+1}^n \subset K_j$$

for each j in $\mathbb{N}$. Let $W_j = \bigcup_{n=1}^{\infty} K_j^n$. Then $K_{j+1} \subset W_{j+1} \subset K_j$ for all j. Thus the uniform structure generated by (W_j) equals the uniform structure generated by (K_j) which is equal to $\mathfrak{U}$.

Finally we will show that $W_j^2 = W_j$ for all j: clearly $W_j \subset W_j^2$. If (x,y) is in W_j^2 there exists a z in X such that $(x,z) \in K_j^n$ and $(z,y) \in K_j^m$ for some (n,m) in $\mathbb{N}^2$. But then (x,y) is in $K_j^{n+m} \subset W_j$. Therefore $W_j^2 = W_j$ and the proof is complete.

<u>Theorem 2.5</u> Let $S \in \mathcal{H}$ be such that $\text{card}(S) \leq \aleph_0$. If $(X,\mathfrak{U})$ is S-metrizable then $(X,\mathfrak{U})$ is $\mathbb{Q}^+$-metrizable.

<u>Proof</u> If S is not dense in any neighborhood of 0, Theorem 2.4 shows that $(X,\mathfrak{U})$ is actually H-metrizable and in particular $\mathbb{Q}^+$-metrizable. If S is dense in a neighborhood of 0, then result is a consequence of Theorem 1.8 and Theorem 1.30.

<u>Theorem 2.6</u> Let $S \in \mathcal{H}$ be a neighborhood of 0 in $\mathbb{R}^+$. If $(X,\mathfrak{U})$ is a metrizable uniform space then $(X,\mathfrak{U})$ is S-metrizable.

Example 2.7 Let $\mathcal{O} = \{S \in \mathcal{H} : S$ is a neighborhood of 0 in $R^+\}$. Then $[S,\mathcal{U},\mathcal{O}]$ is an improper invariant on $\mathcal{U}$ for all S in $\mathcal{O}$.

Presentations and metrization theorems

For the following group of three definitions we will suspend the conventions made at the beginning of this chapter. Let $\mathcal{A}$ be any real-generated category with datum V , let $\mathcal{S} \subset \mathcal{J}$ and let $P(V,\mathcal{S})$ be endowed with the partial ordering defined in Chapter I above. Then a partial presentation of $P(V,\mathcal{S})$ will be a list of some or all of the classes $[S]$, with an explicit $S \in \mathcal{S}$ chosen to represent the class, together with a list of the inequalities between these classes. A presentation of $P(V,\mathcal{S})$ will be a list of some or all of the classes $[S]$, any two of the listed classes being distinct, together with a list of all the inequalities between the listed classes. A complete presentation of $P(V,\mathcal{S})$ will be a presentation for which each class is distinct and no other equivalence classes exist.

Example 2.9 Let $\mathcal{S} = \{\{0\}, I\}$ where $I = [0, 1]$. Then, if $[S] = [S,\mathcal{U},\mathcal{S}]$, $[\{0\}] < [I]$ is a complete presentation for $P(\mathcal{U}, \mathcal{S})$, where $\mathcal{U}$ is the category of metrizable uniform spaces.

Theorem 2.10 Let $\mathcal{S}_1 \subset \mathcal{S}_2 \subset \mathcal{J}$ and let $\mathcal{A}$ be any real-generated category with datum V . Define a map

$$j : P(V, \mathcal{S}_1) \to P(V, \mathcal{S}_2)$$
$$[S, V, \mathcal{S}_1] \mapsto [S, V, \mathcal{S}_2]$$

for each S in $\mathcal{S}_1$. Then $x < y$ if and only if $j(x) < j(y)$ and j is a monomorphism in the category of partially ordered sets.

For each $S \in \mathcal{S}_1$ we may identify $P(V, \mathcal{S}_1)$ with its image in $P(V, \mathcal{S}_2)$ under the map j defined as in Theorem 2.10. In some cases a presentation for $P(V, \mathcal{S}_2)$ will enable us to obtain a complete presentation for $P(V, \mathcal{S}_1)$. We will see several instances of this phenomena below.

Theorem 2.11 Let $[S] = [S, \mathcal{U}, \mathcal{H}]$ for each S in $\mathcal{H}$ Then

$$[\{0\}] < [\{0,1\}] < [H] < [Q^+] < [I]$$

is a presentation for $P(\mathcal{U}, \mathcal{H})$. Furthermore, if $S \in \mathcal{H}$ and [S] is not equal to $[\{0\}]$, $[\{0,1\}]$, [H] or $[Q^+]$ then $[Q^+] < [S] < [I]$.

Proof Because $\{0\} \subset \{0,1\} \subset H \subset Q^+ \cap I \subset I$ it follows that

$$[\{0\}] < [\{0,1\}] < [H] < [Q^+] < [I] .$$

If S is in $\mathcal{H}$ then $[\{0\}] < [S] < [I]$ as we have seen. We will show that if $[S]$ is not equal to one of the first three classes then necessarily $[Q^+] < [S] < [I]$. This will complete the proof.

(a) If $[\{0\}] \neq [S]$, then $[\{0,1\}] < [S]$: Because $\{0\} \neq S$ there exists a $\delta > 0$ such that $\{0,\delta\} \subset S$. But then $\hom(\{0,1\}, S) \neq \emptyset$ and therefore $[\{0,1\}] < [S]$.

(b) If $[\{0,1\}] < [S]$ and $[\{0,1\} \neq [S]$, then $[H] < [S]$: By Theorem 2.1 $0 \in cl(S - \{0\})$ and so there exists a sequence (a_n) in S such that $a_n \downarrow 0$ and $a_n \neq 0$ for all n. If (X,d) is an H-metric space there exists a base $(U_n)_{n \in \mathbb{N}}$ for $\mathfrak{U}_d$ satisfying $U_{n+1} \subset U_n$ and $U_n^2 = U_n$ for all n, by the proof of Theorem 2.4. Let $h(x,y) = \min \{a_n : (x,y) \in U_n\}$. Then h is an S-metric uniformly equivalent to d. Therefore $[H] < [S]$.

(c) If $[H] < [S]$ and $[H] \neq [S]$ then $[Q^+] < [S]$: By Theorem 2.4 $cl(S)$ is a neighborhood of 0. Therefore there exists a $\delta > 0$ such that S is dense in the interval $[0,\delta]$. Hence, by Theorem 1.30,

$$\hom(\mathbb{Q}^+ \cap [0,\delta] , S \cap [0,\delta]) \neq \emptyset$$

and so $[\mathbb{Q}^+] \prec [S]$. This completes the proof.

Notation: we will sometimes write 0 instead of $\{0\}$ and D instead of $\{0,1\}$.

<u>Theorem 2.12</u> Let $\mathcal{S}$ be the family of all at most countable sets in $\mathcal{H}$ and let $[S] = [S,\mathcal{U},\mathcal{S}]$ for each S in $\mathcal{S}$. Then

$$[0] \prec [D] \prec [H] \prec [\mathbb{Q}^+]$$

is a complete presentation for $P(\mathcal{U},\mathcal{S})$.

<u>Proof</u> Let $j : P(\mathcal{U},\mathcal{S}) \to P(\mathcal{U},\mathcal{H})$ be the morphism defined in Theorem 2.10. Let S be in $\mathcal{S}$. By Theorem 2.5

$$[\{0\},\mathcal{U},\mathcal{H}] \prec j([S,\mathcal{U},\mathcal{S}]) \prec [\mathbb{Q}^+,\mathcal{U},\mathcal{H}] .$$

Thus, by Theorem 2.11, $j([S,\mathcal{U},\mathcal{S}])$ must be $[\{0\},\mathcal{U},\mathcal{H}]$, $[\{0,1\},\mathcal{U},\mathcal{H}]$, $[H,\mathcal{U},\mathcal{H}]$ or $[\mathbb{Q}^+,\mathcal{U},\mathcal{H}]$. Because each set $\{0\}$, $\{0,1\}$, H and $\mathbb{Q}^+$ is at most countable, j maps $P(\mathcal{U},\mathcal{S})$ onto the sub partially ordered set of $P(\mathcal{U},\mathcal{H})$ generated by these four elements. Then, by Theorem 2.10, $P(\mathcal{U},\mathcal{S})$ is isomorphic to this sub partially ordered set and

$$[0] \prec [D] \prec [H] \prec [\mathbb{Q}^+]$$

is a complete presentation for $P(\mathcal{U}, \mathcal{S})$.

Theorem 2.13 Let $\mathcal{S}$ be the family of sets S in $\mathcal{H}$ with dimension zero. Let $[S] = [S, \mathcal{U}, \mathcal{S}]$ for each S in $\mathcal{S}$. Then

$$[0] \prec [D] \prec [H] \prec [\mathbb{Q}^+] \prec [\mathbb{P}^+]$$

is a partial presentation for $P(\mathcal{U}, \mathcal{S})$ and $[\mathbb{P}^+]$ is the maximum element.

Proof Firstly, note that each representative listed has dimension zero and that

$$\{0\} \subset \{0,1\} \subset H \subset \mathbb{Q}^+ \subset \pi . \mathbb{P}^+ .$$

Then, by Theorem 1.8 and example 1.10,

$$[0] \prec [D] \prec [H] \prec [\mathbb{Q}^+] \prec [\mathbb{P}^+] .$$

Let $\mathcal{S}_1$ be the at most countable sets in $\mathcal{S}$ and let

$$j: P(\mathcal{U}, \mathcal{S}_1) \to P(\mathcal{U}, \mathcal{S})$$

be the monomorphism defined above. If $S \in \mathcal{S}$ is at most denumerable then [S] belongs to the range of j and must therefore be one of the classes $[\{0\}]$, $[\{0,1\}]$, [H] or $[\mathbb{Q}^+]$. If S in $\mathcal{S}$ is not countable and not dense in any neighborhood of 0 then $[S] \prec [H]$ by Theorem 2.4. Because

$$P(\mathcal{U},\mathcal{S}) \to P(\mathcal{U},\mathcal{H})$$

is a monomorphism, $[S]$ must be $[\{0\}]$, $[\{1,0\}]$ or $[H]$. If S is not countable and is dense in a neighborhood of zero then there is a $\delta > 0$ such that $\hom(S \cap [0,\delta], \mathbb{P}^+) \neq \emptyset$ by Theorem 1.32 and Theorem 1.33. Therefore $[S] \prec [\mathbb{P}^+]$ and $[\mathbb{P}^+]$ is the maximum class.

Theorem 2.14 Let $\mathcal{X}$ be the closed subsets in $\mathcal{H}$ and let $[S] = [S,\mathcal{U},\mathcal{X}]$ for each S in $\mathcal{X}$. Then

$$[0] \prec [D] \prec [H] \prec [I]$$

is a complete presentation for $P(\mathcal{U},\mathcal{X})$.

Proof Observe that each of the listed representatives is closed, that $\{0\} \subset \{0,1\} \subset H \subset I$ and that, by Theorem 2.11, $[\{0\},\mathcal{U},\mathcal{H}] \prec [\{0,1\},\mathcal{U},\mathcal{H}] \prec [H,\mathcal{U},\mathcal{H}] \prec [I,\mathcal{U},\mathcal{H}]$. We will show that the range of the monomorphism

$$j\colon P(\mathcal{U},\mathcal{X}) \to P(\mathcal{U},\mathcal{H})$$

is exactly this sub partially ordered set. The conclusion of the theorem follows from this fact.

If $S \in \mathcal{X}$ is a neighborhood of 0 then $[S,\mathcal{U},\mathcal{H}]$ is equal to $[I,\mathcal{U},\mathcal{H}]$. If S is in $\mathcal{X}$ and is not a neighborhood of 0 then, by Theorem 2.4,

$$[S, \mathcal{U}, \mathcal{H}] < [H, \mathcal{U}, \mathcal{H}] .$$

Therefore $[S, \mathcal{U}, \mathcal{H}]$ is either $[\{0\}, \mathcal{U}, \mathcal{H}]$, $[\{0,1\}, \mathcal{U}, \mathcal{H}]$ or $[H, \mathcal{U}, \mathcal{H}]$. This completes the proof.

Let $\mathcal{A}$ and $\mathcal{B}$ be real-generated categories with datums $V_{\mathcal{A}}$ and $V_{\mathcal{B}}$ respectively. Let $\mathcal{S} \subset \mathcal{J}$ be a non empty subfamily. First suppose that $\mathcal{B}$ is derived from $\mathcal{A}$. Define a map

$$k: P(V_{\mathcal{A}}, \mathcal{S}) \rightarrow P(V_{\mathcal{B}}, \mathcal{S}) \text{ by}$$

$$[S, V_{\mathcal{A}}, \mathcal{S}] \mapsto [S, V_{\mathcal{B}}, \mathcal{S}] \quad \text{for } S \text{ in } \mathcal{S} .$$

It follows from the proof of Theorem 1.9 that if $x < y$ then $k(x) < k(y)$. The map k is a surjective homomorphism in the category of partially ordered sets.

Now suppose that $\mathcal{B} \subset \mathcal{A}$ is a subcategory. The map

$$\ell: P(V_{\mathcal{A}}, \mathcal{S}) \rightarrow P(V_{\mathcal{B}}, \mathcal{S}) \quad \text{defined by}$$

$$[S, V_{\mathcal{A}}, \mathcal{S}] \rightarrow [S, V_{\mathcal{B}}, \mathcal{S}] \quad \text{for } S \text{ in } \mathcal{S}$$

is again a surjective homomorphism in the category of partially ordered sets.

<u>Theorem 2.15</u> Let $\mathcal{C}$ be the class of compact metrizable uniform spaces and, for each S in $\mathcal{H}$, let $[S] = [S, \mathcal{C}, \mathcal{H}]$. Then

$$[0] < [D] < [H] < [I]$$

is a complete presentation for $P(\mathcal{C}, \mathcal{H})$.

<u>Proof</u> The map $\ell : P(\mathcal{U}, \mathcal{H}) \to P(\mathcal{C}, \mathcal{H})$ discussed in the preceding paragraph is a homomorphism and therefore

$$[0] \prec [D] \prec [H] \prec [I]$$

because the same relations hold in $P(\mathcal{U}, \mathcal{H})$. (One may also prove this directly.) To see that the classes are distinct consider the compact subspaces $\{0\}$, $\{1,0\}$, H and I of R each with the metric inherited from R. The existence of the homomorphism ℓ and the corresponding fact for $P(\mathcal{U}, \mathcal{H})$ implies that there are no classes between $[0]$ and $[D]$ and between $[D]$ and $[H]$. Similarly if $[S] \neq [0]$ or $[D]$ then $[H] < [S] < [I]$. If S is in $\mathcal{O}$ then $[S] = [I]$ by Theorem 2.6. If S is not in $\mathcal{O}$ let $(X, \mathfrak{U})$ be such that X is compact. Let d be a metric on X satisfying $d(X^2) \subset S$ and $\mathfrak{U} = \mathfrak{U}_d$. The set $d(X^2)$ is necessarily a closed subset of S and therefore, by Theorem 2.4, $(X, \mathfrak{U})$ is H-metrizable. Thus $[H] = [S]$ and the proof is complete.

Up to this point in the discussion various classes $[S]$ have been examined and many shown to be the same using uniform space constructions or morphisms. A list giving no two classes which have been proved to be alike reads:

$$[0] \prec [D] \prec [H] \prec [\mathbb{Q}^+] \prec [\mathbb{P}^+] \prec \ldots \prec [J_{m+1}] \prec [J_m] \prec \ldots \prec [J_1] \prec [I] .$$

(We have not proved, however, that the set of classes is linearly ordered.)

Although morphisms can sometimes be used to examine the gaps between classes, sometimes they cannot. Under very mild restrictions on S all the functions in $\hom(S,T)$, by Theorem 1.34, are uniformly continuous. This is a severe restriction, clearly, on the type of function that can be employed.

<u>Exercise</u>: Find a subset $S \in \mathcal{H}$ such that $[S,\mathcal{U},\mathcal{H}] = [H,\mathcal{U},\mathcal{H}]$ but $\hom(S,T) = \phi$.

<u>Problem 1</u> If S and T in $\mathcal{H}$ are such that

$$[\mathbb{Q}^+,\mathcal{U},\mathcal{H}] \prec [S,\mathcal{U},\mathcal{H}] \prec [T,\mathcal{U},\mathcal{H}]$$

is $\hom(S,T) \neq \phi$?

<u>Remarks 2.16</u> Let $(X,\mathcal{U})$ be a separated uniform space and for each $S \in \mathcal{H}$ let $[S] = [S,\mathcal{U},\mathcal{H}]$. Then

(a) $(X,\mathcal{U})$ is [O]-metrizable if and only if $\Delta = \mathcal{U}$,

(b) $(X,\mathcal{U})$ is [D]-metrizable if and only if $\Delta \in \mathcal{U}$, by Theorem 2.1,

(c) $(X,\mathcal{U})$ is [H]-metrizable if and only if there exists a base $(U_n)_{n\in\mathbb{N}}$ for $\mathcal{U}$ satisfying $U_n^2 = U_n$ for all n by Theorem 2.2 and the proof of Theorem 2.4,

(d) $(X,\mathcal{U})$ is [I]-metrizable if and only if there exists a denumerable base $(U_n)_{n\in\mathbb{N}}$ for $\mathcal{U}$ by Theorem 2.6 and Bourbaki [11, Chapter IX, §2.4, Theorem 1].

Constructing a statement analagous to these four and corresponding to the property $[\mathbb{Q}^+]$-metrizable uniform space, is not such an easy task. By Theorem 2.11 this property, in a sense, is next in order of complexity. We consider this property below.

Let $(X,\mathcal{U})$ be a uniform space and let $\mathcal{B} = (B_a)_{a\in A}$ be a base for $\mathcal{U}$. The base induces an equivalence relation $\sim$ on X^2 defined by the rule $(x,y) \sim (u,v)$ if and only if, for all a in A , (x,y) is in B_a if and only if (u,v) is in B_a .

Definition 2.17 This equivalence relation induces a partition of X^2 and the cardinality of this partition is defined to be the density of the uniformity base $\mathcal{B}$. We will usually denote this cardinal number by $\# \mathcal{B}$.

Theorem 2.18 Let the separated uniform space $(X,\mathfrak{U})$ admit a base

$$\mathcal{B} = (U(n,m) : (n,m) \in \mathbb{N}^2)$$

of symmetric entourages satisfying

UN 1 $U(1,n) \circ U(m,n) \subset U(m+1, n)$ for all (m,n) in $\mathbb{N}^2$,

UN 2 $U(m,n) = U(2m, n+1)$ for all (m,n) in $\mathbb{N}^2$, and

UN 3 $\# \mathcal{B} \leq \aleph_0$.

Then $(X,\mathfrak{U})$ is $\mathcal{Q}^+$- metrizable. Conversely, if $(X,\mathfrak{U})$ is $\mathcal{Q}^+$-metrizable then there exists a base $\mathcal{B}$ for $\mathfrak{U}$ satisfying UN 1, UN 2, and UN 3.

Proof (sufficiency) By UN 1 and induction we see that $U(r,n) \circ U(s,n) \subset U(r+s, n)$ for all r, s and n in $\mathbb{N}$. Similarly UN 2 implies $U(m,n) = U(2^s m, n+s)$ and $U(r,s) = U(2^n r, s+n)$. Therefore

$U(m,n) \circ U(r,s) \subset U(2^s m+2^n r, n+s)$ for all m, n, r and s .

If (x,y) is in X^2 let

Then, because each $U(m,n)$ is symmetric, $d(x,y) = d(y,x)$ and, by the last formula of the preceding paragraph, d satisfies $d(x,y) \leq d(x,y) + d(z,y)$ for all x, y, and z in X. If $(x,y) \in U(m,n)$ then $d(x,y) \leq m/2^n$ and therefore

$$U(m,n) \subset \{(x,y): d(x,y) \leq m/2^n\} .$$

Let $d(x,y) < m/2^n$. Then there exist integers r and s such that $(x,y) \in U(r,s)$ and $r/2^s \leq m/2^n$. We consider three cases:

(a) If $n = s$: then $r \leq m$ and by UN 1 $U(r,n) \subset U(m,n)$. Therefore (x,y) is in $U(m,n)$.

(b) If $s > n$: then $r \leq m2^{s-n}$ and (x,y) is in $U(r,s)$, which is contained in $U(m2^{s-n}, s)$ by UN 1, which is contained in $U(m, s-s+n)$ by UN 2. Therefore (x,y) is in $U(m,n)$.

(c) If $n > s$: then $r2^{n-s} \leq m$ and $(x,y) \in U(r,s)$ which is equal to $U(2^{n-s}.r, s+n-s)$ by UN 2 which is contained in $U(m,n)$ by UN 1. Therefore

$$\{(x,y): d(x,y) < m/2^n\} \subset U(m,n) \subset \{(x,y): d(x,y) \leq m/2^n\} .$$

Therefore d generates the uniformity $\mathfrak{U}$.

Finally we will prove that d has at most a countable set of values. It follows from the definition of d that if $(x,y) \sim (a,b)$ then $d(x,y) = d(a,b)$. By UN 3 there are at most a countable number of equivalence classes and thus an at most countable set of values for the metric d on X. The conclusion of this part of the theorem then follows by Theorem 2.12.

(necessity) Let $(X,\mathfrak{U})$ be $\mathbb{Q}^+$- metrizable. Then by Theorem 1.30 there exists a metric d on X with values in the dyadic rationals satisfying $\mathfrak{U} = \mathfrak{U}_d$. For (m,n) in $\mathbb{N}^2$ let

$$U(m,n) = \{(x,y): d(x,y) \leq m/2^n\} .$$

Then $\mathcal{B} = (U(m,n): (m,n) \in N^2)$ is a base for $\mathfrak{U}$, each $U(m,n)$ is symmetric, and UN 1 and UN 2 are clearly satisfied. Finally $(x,y) \sim (a,b)$ if and only if $d(x,y) = d(a,b)$ and therefore $\# \mathcal{B} \leq \aleph_0$. This completes the proof.

<u>Problem 2</u> Let (X,d) be a J_1-metric space where induced topology is separable. Does there exist a $\mathbb{Q}^+$- metric on X which is uniformly equivalent to d ?

<u>Problem 3</u> Let $X \subseteq R$ be the classical unmeasurable set and let d be the usual metric on R. Then $d(X^2) \subseteq \mathbb{P}^+$. Is it true that no metric on X, uniformly equivalent to d on X, has values in $\mathbb{Q}^+$?

Example 2.19 Let $S = \mathbb{Q}^+ \bigcup_{n=1}^{\infty} [\frac{1}{2^{2n+1}}, \frac{1}{2^{2n}}]$.

Let $a_n = 2^{-2n}$, $b_n = 2^{-2n-1}$ and c_n be defined as in Theorem 1.31. Then $c_n \to 0$ and thus $[S] = [\mathbb{Q}^+]$, by Theorem 1.9, whereas S is a "large" set.

Theorem 2.20 If S has a countable complement then $[\mathbb{P}^+] < [S]$ where $\mathbb{P}$ stands for the irrational numbers.

Proof The sets $\mathbb{Q}^+ \setminus \{0\}$ and $I \setminus S$ satisfy the hypotheses of Theorem 1.33 and therefore, by Theorem 1.32 $\hom(\mathbb{P}^+, S) \neq \phi$. Thus $[\mathbb{P}^+] < [S]$.

We will now define a new datum for $\mathcal{U}$, the category of metrizable uniform spaces.

Definition 2.21 If $(X, \mathfrak{U})$ is in $\mathcal{U}$ the uniform structure $\mathfrak{U}$ will be generated by a function $d: X^2 \to \mathbb{R}$ satisfying

(1) $d(x,y) = 0$ if and only if $x = y$,

(2) $d(x,y) \geq 0$ for all x and y,

(3) $d(x,y) = d(y,x)$ for all x and y,

(4) there exists an r in $(0,1]$ such that

$r.d(x,y) \leq d(x,z) + d(y,z)$ for all x, y and z,

(5) d is a continuous with respect to the topology $\mathcal{T}_{\mathfrak{U}} \times \mathcal{T}_{\mathfrak{U}}$.

We call such a function a continuous r-metric.

Let $\mathcal{P}_\alpha$ be the conjunction of these five properties. Let $\mathcal{R}_\alpha$ be the rule: U is in $\mathfrak{U}$ if and only if there exists $\epsilon > 0$ such that $\{(x,y): d(x,y) < \epsilon\} \subset U$. Let $V_\alpha = (\mathcal{U}, \mathcal{P}_\alpha, \mathcal{R}_\alpha)$.

<u>Theorem 2.22</u> For each S in $\mathcal{H}$ let $[S] = [S, V_\alpha, \mathcal{H}]$. Then

$$[0] \prec [D] \prec [H] \prec [I]$$

is a complete presentation for $P(V_\alpha, \mathcal{H})$.

This theorem follows, almost directly, from Theorem 2.23 below.

Let $X_1 = \{2^{-n}: n \in \mathbb{N}\} \cup \{0\}$. (The significance of the subscript will appear in Chapter III.)

<u>Theorem 2.23</u> Let $b_n \downarrow 0$ be a sequence of distinct real numbers and let $S = [0,1] \setminus \{b_n\}$. Let (X,d) be an S-metric space. There is a continuous $X_1 - \frac{1}{2}$ - pseudometric on X generating the same uniform structure as d.

<u>Proof</u> Let $n_1 = 1$ and choose n_{i+1} by induction so that $b_{n_{i+1}} < \frac{b_{n_i}}{3}$. Let $\Delta = \{(x,x): x \in X\}$ and

$K_i = \{(x,y): b_{n_{i+1}} < d(x,y) < b_{n_i}\}$ for each i in $\mathbb{N}$.

Let $H_i = \Delta \cup \bigcup_{j \geq i} K_j$, $K_o = \{(x,y): d(x,y) > b_{n_1}\}$, and $H_o = X \times X$. Then each K_i and each H_i is a closed and open set in $X \times X$. The family of sets $(H_i)_{i \geq 0}$ is a uniformity base and generates a uniform structure equivalent to that generated by the metric d. Let χ_{K_i} be the characteristic function of K_i and define

$$a_1(x,y) = \sum_1^\infty \frac{1}{2^i} \chi_{K_i}(x,y)$$

for each (x,y) in $X \times X$. Then a_1 is a continuous function. If x, y, and z are in X then

$$a_1(x,y) \leq 2 \max \{a_1(x,z), a_1(z,y)\}.$$

This implies a_1 is a continuous $X_1 - \frac{1}{2}$ - pseudometric on X. The sets $F_n = \{(x,y): a_1(x,y) < \frac{1}{2^n}\}$ generate a uniform structure equivalent to that generated by d. This completes the proof.

Let $\mathcal{Y} = \{S \in \mathcal{H}: S + S \subset S\}$ and let $[S] = [S, \mathcal{U}, \mathcal{Y}]$ for each S in $\mathcal{Y}$.

Notation: let $\mathbb{Z}$ be the integers and $\mathbb{Z}^+ = \mathbb{N} \cup \{0\}$.

<u>Theorem 2.24</u> $[0] \prec [Z^+] \prec [Q^+] \prec [R^+]$ is a presentation for $P(\mathcal{U},\mathcal{Y})$. The first three terms form the initial segment and R^+ is the maximum class. If $S \in \mathcal{Y}$ is such that $[Q^+] \prec [S]$ and $[Q^+] \neq [S]$ then S is uncountable.

<u>Proof</u> Because $\{0\} \subset Z^+ \subset Q^+ \subset R^+$ we see that $[0] \prec [Z^+] \prec [Q^+] \prec [R^+]$ by Example 1.10. By Example 1.13, R^+ is the maximum class. The mapping

$$P(\mathcal{U},\mathcal{Y}) \to P(\mathcal{U},\mathcal{H}) \quad \text{defined by}$$

$$[S,\mathcal{U},\mathcal{Y}] \to [S,\mathcal{U},\mathcal{H}] \quad \text{for } S \text{ in } \mathcal{Y}$$

is an injective homomorphism. It follows that the given classes are distinct because, by Theorem 2.11, the classes $[0,\mathcal{U},\mathcal{H}]$, $[Z^+,\mathcal{U},\mathcal{H}]$, $[Q^+,\mathcal{U},\mathcal{H}]$, and $[R^+,\mathcal{U},\mathcal{H}]$ are distinct.

(a) If $S \in \mathcal{Y}$ is such that $[0] \prec [S]$ and $[0] \neq [S]$ then $[Z^+] \prec [S]$ because $[Z^+,\mathcal{U},\mathcal{H}] = [D,\mathcal{U},\mathcal{H}]$ and the result is true in $P(\mathcal{U},\mathcal{H})$.

(b) If $S \in \mathcal{Y}$ is such that $[Z^+] \prec [S]$ and $[Z^+] \neq [S]$ then $[Q^+] \prec [S]$: by Theorem 2.11 again we see that necessarily $[H,\mathcal{U},\mathcal{H}] \prec [S,\mathcal{U},\mathcal{H}]$. But then $[H,\mathcal{U},\mathcal{H}] = [S,\mathcal{U},\mathcal{H}]$ or $[Q^+,\mathcal{U},\mathcal{H}] \prec [S,\mathcal{U},\mathcal{H}]$. If the latter is true then $[Q^+] \prec [S]$ and we are done. If the former is true then S is necessarily not dense in a neighborhood of zero in $[0,\infty)$ and there exists a sequence of points tending to zero in S. This is impossible because $S + S \subset S$ and therefore this case does not occur.

(c) Let $[\mathbb{Q}^+] < [S]$. If $[\mathbb{Q}^+] \neq [S]$ then $[\mathbb{Q}^+, \mathcal{U}, \mathcal{H}]$ is not equal to $[S, \mathcal{U}, \mathcal{H}]$ and therefore S is uncountable by Theorem 1.30. This completes the proof.

Let $\mathcal{W} = \{S \in \mathcal{Y} : S = G^+ = G \cap [0,\infty)$ where G is an additive subgroup of $R\}$.

<u>Theorem 2.25</u> Let $[S] = [S, \mathcal{U}, \mathcal{W}]$ for each $S \in \mathcal{W}$. Then

$$[0] < [\mathbb{Z}^+] < [\mathbb{Q}^+] < [R^+]$$

is a presentation for $P(\mathcal{U}, \mathcal{W})$ in which the first three terms form the initial segment. If $S \in \mathcal{W}$ is such that $[S] = [\mathbb{Z}^+]$ then $S = a.\mathbb{Z}^+$ for some $a > 0$ in R. If $S \in \mathcal{W}$ and $S \neq R^+$ then $[S, \mathcal{U}, \mathcal{H}] < [\mathbb{P}^+, \mathcal{U}, \mathcal{H}]$.

<u>Proof</u> (a) The presentation follows readily from Theorem 2.24 and the mapping

$$[S, \mathcal{U}, \mathcal{W}] \to [S, \mathcal{U}, \mathcal{Y}], \quad S \in \mathcal{W}.$$

(b) If $S \in \mathcal{W}$ satisfies $[S] = [\mathbb{Z}^+]$ then, because the class $[S, \mathcal{U}, \mathcal{H}] = [D, \mathcal{U}, \mathcal{H}]$, we have $0 < \text{glb}\{S - \{0\}\} = a$ say. Let $G = S - S$. Then necessarily 0 is an isolated point of the subgroup G, and therefore each point in G is isolated. Hence $a \in G$ and $G = a.\mathbb{Z}$. Therefore $S = a.\mathbb{Z}^+$.

(c) If $b \in R^+ - S$ then $\{b/n: n \in \mathbb{N}\} \subset R^+ - S$. Thus the connected component containing 0 in S is $\{0\}$ by Bourbaki [10, Chapter III, Proposition 2.2.7] and hence, S does not contain an interval. Therefore, by Theorem 2.13, $[S,U,H] < [\mathbb{P}^+,U,H]$.

<u>Theorem 2.26</u> Let $\mathcal{S} \subset \mathcal{W}$ be the family of positive parts of closed additive subgroups of R. If $[S] = [S,\mathcal{U},\mathcal{S}]$ for each S in $\mathcal{S}$, then

$$[0] < [\mathbb{Z}^+] < [R^+]$$

is a complete presentation for $P(\mathcal{U},\mathcal{S})$.

<u>Proof</u> Let $\mathcal{Z}$ be the family of closed subsets in $\mathcal{H}$. The result follows by considering the mapping

$$[S,\mathcal{U},\mathcal{W}] \to [S,\mathcal{U},\mathcal{Z}]$$

and then applying Theorem 2.14.

<u>Remark</u> After observing that $[S,\mathcal{U},\mathcal{S}] = [R^+,\mathcal{U},\mathcal{S}]$ if and only if $S = R^+$ we conclude that the only proper closed subgroups of R are of the form $a.\mathbb{Z}$ for some $a > 0$ in R, Bourbaki [11, Chapter V, Proposition 1.1.1].

Uniform Spaces

Let $\mathcal{E}$ be the class of all uniform spaces. If X is a set and $F: X^2 \to [0,\infty)$ a family of pseudometrics let $\mathfrak{U}_F$ be the uniform structure on X generated by the sets $\{A \subset X^2$: there exists $d_1, \ldots, d_n$ in F and $\epsilon > 0$ such that if $d_i(x,y) < \epsilon$ for $1 \leq i \leq n$ then $(x,y) \in A\}$. If $(X,\mathfrak{U}) \in \mathcal{E}$ then there exists a family $F: X^2 \to [0,\infty)$ satisfying $\mathfrak{U} = \mathfrak{U}_F$. In this case $\mathcal{P}$ is the property, "F is a set of pseudometrics on X" and the rule $\mathcal{R}$ is as we have just described it.

Theorem 2.7 Let $V = (\mathcal{E},\mathcal{P},\mathcal{R})$ be the datum of $\mathcal{E}$ and let $[S] = [S,V,\mathcal{H}]$ for each S in $\mathcal{H}$. Then

$$[0] \prec [D] \prec [Q^+] \prec [I]$$

is a presentation for $P(V,\mathcal{H})$ and the first three terms form the initial segment.

Proof (a) $S \in \mathcal{H}$ then $[0] \prec [S] \prec [I]$: clearly $0 \prec S$. If $(X,\mathfrak{U})$ is a uniform space and $F: X^2 \to S$ satisfies $\mathfrak{U} = \mathfrak{U}_F$ let $G = \{\min\{1,d\}: d \in F\}$. Then $\mathfrak{U} = \mathfrak{U}_G$ and thus $S \prec I$.

(b) If $[0] \prec [S]$ and $[0] \neq [S]$ then $[D] \prec [S]$:

there exists an $a > 0$ such that $\{0,a\} \subset S$. If $(X,\mathfrak{U})$ is a space and $F: X^2 \to D$ satisfies $\mathfrak{U} = \mathfrak{U}_F$ then $G = \{a.d: d \in F\}$ satisfies $\mathfrak{U} = \mathfrak{U}_F$ and $G: X^2 \to S$. Thus $[D] < [S]$.

(c) If $[D] < [S]$ and $cl(S)$ is not a neighborhood of 0 in R^+ then $[D] = [S]$: let $(X,\mathfrak{U})$ be a space and $F: X^2 \to S$ a family of pseudometrics with $\mathfrak{U} = \mathfrak{U}_F$. Let $d \in F$. Then there exists a pseudometric h_d on X with $h_d: X^2 \to W$ and $(X,\mathfrak{U}_d) = (X,\mathfrak{U}_{h_d})$ by Theorem 2.2 and Theorem 2.4. Let $n \in \mathbb{N}$ and set $U(d,n) = \{(x,y) \in X^2: h_d(x,y) \leq 3^{-n}\}$. Then $U(d,n) \circ U(d,n) = U(d,n)$. Define a map $k(d,n)$ on X^2 by

$$k(d,n)\,(x,y) = \begin{cases} 0 & \text{if } (x,y) \in U(d,n) \\ 1 & \text{otherwise .} \end{cases}$$

Then the family $\{k(d,n): d \in F , n \in \mathbb{N}\} = K$ is a set of pseudometrics on X , $K: X^2 \to D$ and $\mathfrak{U}_K = \mathfrak{U}_F = \mathfrak{U}$. Thus $[S] < [D]$ and therefore $[S] = [D]$.

(d) If S is dense in a neighborhood of 0 in R^+ then $[Q^+] < [S]$: if $F: X^2 \to Q^+$ and $g \in \hom(Q^+, S)$ (Theorem 1.30) let $G = \{g \circ f : f \in F\}$.

(e) $[0] \neq [D]$: consider D with $F = \{d\}$ where $d(0,1) = 1$.

(f) $[D] \neq [Q^+]$: consider the metric $d(x,y) = |x-y|$ on Q. Suppose there exists $F: Q^2 \to D$ with $\mathcal{U}_{\{d\}} = \mathcal{U}_F$. Let $f \in F$ and set $U(f) = \{(x,y): f(x,y) < 1\}$. Then $U(f)^2 = U(f)$. If $\{f_1, \ldots, f_n\} \subset F$ and $V = U(f_1) \cap \ldots \cap U(f_n)$ then $V^2 = V$. Thus $\mathcal{U}$ has a transitive base. Because it has a denumerable base it has a denumerable transitive base. This implies $(Q, \mathcal{U})$ is W-metrizable and that its completion $(R, \hat{\mathcal{U}})$ is W-metrizable -- this is false because R is connected. Therefore no such family F exists and we have proved that $[D] \neq [Q^+]$.

(g) $[Q^+] \neq [I]$: consider $d(x,y) = |x-y|$ on $I = [0,1]$. If $F: I^2 \to Q^+$ satisfies $\mathcal{U}_{\{d\}} = \mathcal{U}_F$ then $\mathcal{T}_{\mathcal{U}_{\{d\}}} = \mathcal{T}_{\mathcal{U}_F}$. This is impossible because $\mathcal{T}_{\mathcal{U}_F}$ necessarily has dimension zero. This completes the proof of the theorem.

<u>Remark 2.28</u> It is possible to study the family of quasi-uniform spaces in a similar manner. In this case the family of functions will be quasi-metrics. The appropriate theory of quasi-uniform spaces, including a rule for the generation of quasi-uniform structures, may be found in Reilly [53].

Chapter III : Topological Spaces

S-metrizable topological spaces

A proof similar to that of [15, Theorem 3.1] will demonstrate

Theorem 3.1 Let $S \in \mathcal{H}$ be such that S is not a neighborhood of zero in R^+. If the topological space $(X,\mathcal{T})$ is S-metrizable then $\operatorname{ind} X = 0$.

Theorem 3.2 The Hausdorff space $(X,\mathcal{T})$ is metrizable and has $\operatorname{ind} X = 0$ if and only if it is paracompact and admits a clopen development.

Proof (necessity) Let d be a metric on X generating the topology $\mathcal{T}$ and, for each n in $\mathbb{N}$, let

$$B(x,1/n) = \{y \in X : d(x,y) < 1/n\}$$

be the metric ball with center x. For each point x in X and integer n choose a clopen set G with $x \in G \subset B(x,1/n)$. Let G_n be the family of all such sets G where n is fixed. Then $(G_n)_{n \in \mathbb{N}}$ is a clopen development for X. Because X is metrizable it is paracompact [64].

(sufficiency) If X is paracompact and admits a clopen development it is a regular paracompact Moore space, and so, by [7] and [43], X is metrizable. Clearly $\text{ind}\, X = 0$ also.

There are subclasses of the class of metrizable spaces for which we can prove a converse for Theorem 3.1 as will be seen in the sequel. However the general case is obscure.

The following theorems have been proven by the author:

Theorem [13, Corollary following Theorem 2] The space $(X,\mathcal{T})$ has Čech (large inductive) dimension zero if and only if there exists a metric for X generating the topology $\mathcal{T}$, having values in R^+ with zero as the only cluster point of those values.

Theorem [14, Corollary 2] A metrizable space has a compatible metric taking values in a closed subset of R^+ having dimension zero if and only if the space has Čech dimension zero.

Definition [14, page 116] Let $(X,\mathcal{T})$ be a topological space and let $D = (D_n)_{n\in\mathbb{N}}$ be a development for $\mathcal{T}$. If the sets in each open cover D_n are disjoint we say that $(X,\mathcal{T})$ is sievable and call the family of covers D a compatible sieve for X .

<u>Theorem</u> [14, Theorem 4] Let $(X,\mathcal{T})$ be a T_2 topological space. Then $(X,\mathcal{T})$ is sievable if and only if it is metrizable and has Čech dimension zero.

We will obtain an improvement on these results below following a preliminary theorem.

If $d: X^2 \to S \subset \mathbb{R}$ is a metric on the set X we say (X,d) is an S-metric space.

<u>Theorem 3.3</u> Let $S \in \mathcal{H}$ be such that for some a and b in $\mathbb{R}$ with $0 \leq a < b$ we have $(a,b) \cap S = \phi$. Let (X,d) be an S-metric space and let $b < \epsilon < a$. Then the family of balls $(B(x,\epsilon) : x \in X)$ is closure preserving.

<u>Proof</u> Let $A \subset X$ be a nonempty subset and let

$$y \in \mathrm{cl}\{\bigcup_{x \in A} B(x,\epsilon)\} \setminus \bigcup_{x \in A} B(x,\epsilon) .$$

Then there exist sequences of points (a_n), (x_n) in X with $d(y, a_n) < \frac{1}{n}$ and $d(a_n, x_n) < \epsilon$ for each n in $\mathbb{N}$, the positive integers. Therefore $d(y, x_n) < \epsilon + \frac{1}{n}$ for each n . This implies $d(y, A) \leq \epsilon$ and therefore $d(y, A) \leq a$. However if y is not in $B(x,\epsilon)$ for each

x in A then $d(y, A) \geq \varepsilon$. This contradiction shows that

$$\mathrm{cl}\ \{ \bigcup_{x \in A} B(x, \varepsilon) \} = \bigcup_{x \in A} B(x, \varepsilon) .$$

The result of the theorem follows upon observing that

$$B(x, \varepsilon) = B(x, \varepsilon] = \{y \in X : d(x,y) \leq \varepsilon\}$$

is a closed set.

<u>Theorem 3.4</u> Let $S \in \mathcal{H}$ be not dense in any neighborhood of zero. If $(X, \mathcal{T})$ is S-metrizable then $\mathrm{Ind}\ X = 0$.

<u>Proof</u> Let d be a compatible S-metric on X . Firstly, if S is not dense in any neighborhood of zero, there exists a sequence (c_n) of distinct points of $(0, \infty)$ with $c_1 > c_2 > c_3 \ldots > 0$, $c_n \to 0$, and c_n is not in $\mathrm{cl}(S)$ for each n in $\mathbb{N}$. Choose an open interval neighborhood (a_{n+1}, b_n) for each c_n which is disjoint from S and such that $(a_{n+1}, b_n) \cap (a_{m+1}, b_m) = \phi$ when $n \neq m$. Then $a_1 \geq b_1 > a_2 \geq b_2 \ldots > 0$ and

$$[0, a_1] \cap S \subset \{0\} \cup \bigcup_{n=1}^{\infty} [b_n, a_n] .$$

Fix n in $\mathbb{N}$ and let ε_n be any number with $a_{n+1} < \varepsilon_n < b_n$. Then, by Theorem 3.3, the family of clopen sets $(B(x, \epsilon_n) : x \in X)$ is closure preserving. Well order the points of X and let

$$A(x,n) = B(x, \epsilon_n) \setminus \bigcup_{y < x} B(y, \epsilon_n) .$$

Then each $A(x,n)$ is a clopen (possibly empty) set. If y is the first element of X with $x \in B(y, \epsilon_n)$, then $x \in A(y,n)$. Therefore $(A(x,n) : x \in X)$ is a cover of X of clopen sets. If $x < y$ then $A(x,n) \cap A(y,n) = \phi$ as $A(x,n) \subset B(x, \epsilon_n)$. Therefore the cover $(A(x,n) : x \in X)$ consists of disjoint sets.

We will define a sieve ([13]) for X as follows:

$$\mathcal{D}_1 = (A(x,1) : x \in X) \quad \text{and, if}$$

$$\mathcal{D}_1 , \ldots, \mathcal{D}_n \quad \text{have been defined then}$$

$$\mathcal{D}_{n+1} = (B \cap A(x, \epsilon_{n+1}) : B \in \mathcal{D}_n , x \in X) .$$

Then $(\mathcal{D}_n)$ is a sieve on X and we need only show that it is compatible with the topology induced by the metric d. Let $\delta > o$ be given. Then, because $\epsilon_n \to 0$, there is an n in $\mathbb{N}$ with $\epsilon_n < \delta | 2$. Because $(A(x,n) : x \in X)$ is a cover of X, $x \in A(y,n)$ for some y in X. But then $B(y, \epsilon_n) \subset B(x,\delta)$ and hence $x \in A(y,n) \subset B(x,\delta)$. But $\mathcal{D}_n$ refines $(A(x,n) : x \in X)$ and thus $(\mathcal{D}_n)$ generates the topology $\mathcal{T}$. Hence, by [13, Theorem 4], X has large inductive dimension zero. This completes the proof.

Clearly, H is not dense in any neighborhood of zero, and we deduce immediately:

Theorem 3.5 Let $(X,\mathcal{T})$ be a metrizable space. Then $(X,\mathcal{T})$ has a large inductive dimension zero if and only if there exists a compatible S-metric on X where $S \in \mathcal{H}$ is not dense in any neighborhood of zero.

Presentations

We will adopt the following conventions for the next group of theorems: the rule $\mathcal{R}$ is the usual rule for the generation of a topology by a metric. Let $\mathcal{J}$ be the category of metrizable spaces and let $\mathcal{A} \subset \mathcal{J}$ be a subcategory. If the datum of $\mathcal{A}$ is $V = (\mathcal{A}, \mathcal{P}, \mathcal{R})$ and $\mathcal{S} \subset \mathcal{H}$ is a nonempty subfamily we will write $[S, \mathcal{A}, \mathcal{S}]$ or $[S]$ instead of $[S, V, \mathcal{S}]$ and $P(\mathcal{A}, \mathcal{S})$ instead of $P(V, \mathcal{S})$.

The category $\mathcal{J}$ may be derived from the category $\mathcal{U}$ and so, by the proof of Theorem 1.9, the map

$$k : P(\mathcal{U}, \mathcal{S}) \to P(\mathcal{J}, \mathcal{S})$$

$$[S, \mathcal{U}, \mathcal{S}] \to [S, \mathcal{J}, \mathcal{S}] \quad \text{for } S \text{ in } \mathcal{S} ,$$

is a surjective homomorphism. We will use the map k to

derive presentations for the partially ordered sets $P(\mathcal{T},\mathcal{S})$ from the presentations for sets $P(\mathcal{U},\mathcal{S})$ developed in Chapter II.

<u>Theorem 3.6</u> Let $\mathcal{S}$ be the family of at most countable sets in $\mathbb{N}$ and let $[S] = [S,\mathcal{T},\mathcal{S}]$ for S in $\mathcal{S}$. Then

$$[O] \prec [D] \prec [H] \prec [\mathbb{Q}^+]$$

is a partial presentation for $P(\mathcal{T},\mathcal{S})$. There are no more classes, the first three classes are distinct and form the initial segment and the class $[\mathbb{Q}^+]$ is the maximum.

<u>Proof</u> The image of the complete presentation for $P(\mathcal{U},\mathcal{S})$ given in Theorem 2.12 under the map k (by an abuse of language) is the presentation occuring in the theorem statement. Because k is surjective there are no more equivalence classes. It is easy to check that the first class is not equal to the second and that the second class is not equal to the third.

<u>Theorem 3.7</u> Let $\mathcal{S}$ be the family of sets in having dimension zero and let $[S] = [S,\mathcal{T},\mathcal{S}]$ for each S in $\mathcal{S}$. Then

$$[O] \prec [D] \prec [H] \prec [\mathbb{Q}^+] \prec [\mathbb{P}^+]$$

is a partial presentation for $P(\mathcal{J}, \mathcal{S})$ and $\mathbb{P}^+$ is the maximum class.

<u>Proof</u> This follows from Theorem 2.13.

<u>Theorem 3.8</u> Let $\mathcal{X}$ be the closed sets in $\mathcal{H}$ and let $[S] = [S, \mathcal{J}, \mathcal{X}]$ for each S in $\mathcal{X}$. Then

$$[0] < [D] < [H] < [I]$$

is a complete presentation for $P(\mathcal{J}, \mathcal{X})$.

<u>Proof</u> Apply k to the presentation for $P(\mathcal{U}, \mathcal{X})$ given in Theorem 2.14.

<u>Theorem 3.9</u> Let $[S] = [S, \mathcal{J}, \mathcal{H}]$ for all S in $\mathcal{H}$. Then

$$[0] < [D] < [H] < [Q^+] < [J_1] < [I]$$

is a partial presentation for $P(\mathcal{J}, \mathcal{H})$. The classes $[\{0\}]$ $[\{0,1\}]$, $[H]$, and $[I]$ are distinct. If $[S]$ is not equal to any of these distinct classes then $[Q^+] < [S] < [J_1]$

<u>Problem 4</u> Is $[H, \mathcal{J}, \mathcal{H}] = [Q^+, \mathcal{J}, \mathcal{H}]$?

Remarks 3.10 Let $[S] = [S, \mathcal{J}, \mathcal{H}]$ for S in $\mathcal{H}$.

(i) Let $(X,\mathcal{T})$ be a regular Hausdorff topological space. Then $(X,\mathcal{T})$ is [0]-metrizable if and only if $\mathcal{T}$ is indiscrete. The space $(X,\mathcal{T})$ is [D]-metrizable if and only if $\mathcal{T}$ is discrete. The space $(X,\mathcal{T})$ is [H]-metrizable if and only if there exists a base $\mathcal{B} = (\mathcal{B}_n)_{n\in\mathbb{N}}$ for $\mathcal{T}$ with each family $\mathcal{B}_n$ discrete and $\cup \mathcal{B}_n = X$ for each n , by Broughan [14]. The space $(X,\mathcal{T})$ is [I]-metrizable if and only if there exists a σ-discrete base $(\mathcal{B}_n)_{n\in\mathbb{N}}$ for $\mathcal{T}$, R.H. Bing [7].

(ii) It is possible to show that $[H, \mathcal{A}, \mathcal{H}]$ is not equal to the class $[\mathbb{Q}^+, \mathcal{A}, \mathcal{H}]$ when $\mathcal{A}$ is the category of metrizable uniform spaces with second countable induced topologies and that, when $\mathcal{B}$ is the category of second countable metrizable spaces, the class $[H, \mathcal{B}, \mathcal{H}] = [\mathbb{Q}^+, \mathcal{B}, \mathcal{H}]$. In other words, there exist subcategories $\mathcal{A}$ of $\mathcal{U}$ and $\mathcal{B}$ of $\mathcal{J}$ such that $P(\mathcal{A}, \mathcal{H})$ is not equal to $P(\mathcal{U}, \mathcal{H})$. We may compare this result with the existence of the identities in Theorems 1.22 and 1.23: $P(V_{\mathcal{U}}, \mathcal{H}) = P(V_{\mathcal{n}}, \mathcal{H})$ and $P(V_{\mathcal{J}}, \mathcal{H}) = P(V_{\mathcal{b}}, \mathcal{H})$ respectively.

Consider the category of all metrizable spaces having any property $\mathcal{P}$ which satisfies the following: for all metrizable spaces X, if X satisfies $\mathcal{P}$ and $\operatorname{ind} X = 0$, then $\operatorname{Ind} X = 0$. We will use the special name $\mathcal{J}$ for this category.

<u>Theorem 3.11</u> For each S in $\mathcal{H}$ let $[S] = [S, \mathcal{J}, \mathcal{H}]$. Then

$$[0] < [D] < [H] < [I]$$

is a complete presentation for $P(\mathcal{J}, \mathcal{H})$.

<u>Proof</u> If $(X, \mathcal{T})$ is J_1- metrizable and is in class $\mathcal{J}$ then $\operatorname{ind} X = 0$ by Theorem 3.1 and therefore, by Theorem 3.5, $(X, \mathcal{T})$ is H-metrizable. This proves that $[H] = [J_1]$. The rest of the proof is clear.

In order to understand $\mathcal{J}$ we will study the equivalent notion in the category of topological spaces.

<u>Dimension Zero Topological Spaces</u>

<u>Definition 3.12</u> We call a topological space clopen-paracompact if every clopen cover of the space has a clopen locally finite refinement. We will see that a paracompact space has large inductive dimension zero if and

and only if it has small inductive dimension zero and is clopen-paracompact.

It is well known that the two types of dimension zero are distinct. C.H. Dowker in [20] gave an example of a normal space X with $\operatorname{ind} X = 0$ and $\operatorname{Ind} X \neq 0$. P. Roy has shown that they are distinct even for metrizable spaces. In [57] he gave an example of a topologically complete space X with $\operatorname{ind} X = 0$ and $\operatorname{Ind} X = 1$. We will develop necessary and sufficient conditions for the two concepts of dimension zero to coincide. In [42] K. Morita proved that they were the same for Lindelof spaces. If the spaces considered are metrizable then either of the concepts local compactness [22, page 230, Ex. C] or strong paracompactness [42] is sufficient to ensure that spaces of small inductive dimension zero are also of large inductive dimension zero. The same is true for metrizable spaces having a closed countable covering of strongly paracompact spaces (see K. Morita, [4]). It is not clear that any of these sufficient conditions is necessary even for metrizable spaces. For example if X is an uncountable number of disjoint copies of $\mathbb{Q}$, the rational numbers, then $\operatorname{Ind} X = 0$ but X is not Lindelof. The object of this section is to develop a sufficient condition, for paracompact spaces of small inductive dimension zero to be of large inductive dimension zero, which is also necessary.

Some parts of the work on dimension zero are contained implicitly in the literature, see Ponomarev [50] say. The presentation given here is in a form suitable for the study of classes $[S, \mathcal{C}, \mathcal{H}]$ where $\mathcal{C}$ is a subcategory of $\mathcal{J}$ as evidenced by Theorem 3.11 above.

Let us define the concepts that will be needed subsequently.

Definition 3.13 A space is strongly paracompact (abbreviated to SP) if every open cover has a star finite open refinement. (This concept is well known although I do not know who first introduced it. See [22, page 225].)

A space is clopen-paracompact (CP) if every clopen cover has a clopen locally finite refinement.

A space is strongly clopen-paracompact (SCP) if every clopen cover has a clopen star finite refinement.

A space satisfies property $\mathcal{B}_1$ if every open cover has a clopen locally finite refinement.

A space satisfies property $\mathcal{B}_2$ if every open cover has a clopen σ-locally finite refinement.

A space satisfies property $\mathcal{B}_3$ if every clopen cover has a clopen star countable refinement.

We will verify the following implications for $(X,\mathcal{T})$ a topological space:

$$SP \Rightarrow \mathcal{B}_3 \Rightarrow SCP \Rightarrow CP \underset{\text{ind } X=0}{\Rightarrow} \text{Ind } X = 0$$

$$\text{Ind } X = 0 \underset{X \text{ paracompact}}{\Rightarrow} \mathcal{B}_1 \Rightarrow \mathcal{B}_2 \Rightarrow CP .$$

<u>Lemma 3.14</u> In any topological space every countable clopen cover has a clopen star finite refinement.

<u>Proof</u> Let $(X,\mathcal{T})$ be a topological space and let $(A_n)_{n\in\mathbb{N}}$ be a countable cover of clopen sets. For each n in $\mathbb{N}$ let $B_n = A_n \setminus \bigcup_{j<n} A_j$. Let $x \in X$ and suppose that A_k is the first element of the cover $(A_n)_{n\in\mathbb{N}}$ with the property $x \in A_k$. Then $x \in B_k$. Thus the family $(B_n)_{n\in\mathbb{N}}$ is a clopen cover of X . For fixed k , B_k meets at most k members of the family $(B_n)_{n\in\mathbb{N}}$ as, if $j > k$, $A_k \cap B_j = \phi$ and $B_k \subset A_k$. Therefore $(B_n)_{n\in\mathbb{N}}$ is a clopen star finite refinement of $(A_n)_{n\in\mathbb{N}}$.

Note that the construction used in this lemma is the same as that used in [21, p. 164].

<u>Theorem 3.15</u> Let $(X,\mathcal{T})$ be a strongly paracompact topological space. Then every clopen cover of X has a clopen star countable refinement. That is, if $(X,\mathcal{T})$ is SP, then it is also $\mathcal{B}_3$.

<u>Proof</u> We shall use the equivalent formulation of strongly paracompact spaces developed by Yu. M. Smirnov in [61]. Namely, a space is strongly paracompact, if and only if every open cover has a closed locally finite and star countable refinement.

Let $\mathcal{A} = (A_\lambda)_{\lambda \in \Lambda}$ be a clopen cover of X and let $\mathcal{B}$ be a closed locally finite and star countable refinement of $\mathcal{A}$. The family $\mathcal{B}$ may be partitioned into its components, $\mathcal{B} = \bigcup_{\gamma \in \Gamma} \mathcal{B}_\gamma$. (For the concept of component of a cover see [22].) Each component $\mathcal{B}_\gamma$ is necessarily countable, $\mathcal{B}_\gamma = (B_{\gamma,i})_{i \in \mathbb{N}}$. The sets $C_\gamma = \bigcup_{n=1}^{\infty} B_{\gamma,i}$ are disjoint, and because $\mathcal{B}$ is a locally finite family, they are clopen. (C_γ is closed as it is the union of a locally finite family of closed sets; $X \sim C_\gamma$ is closed for the same reason.)

For each γ in Γ and i in $\mathbb{N}$ let $\lambda(\gamma,i)$ denote a definite element of Λ with the property $B_{\gamma,i} \subset A_{\lambda(\gamma,i)}$. Then the family of sets

$\mathcal{D} = (C_\gamma \cap A_{\lambda(\gamma,i)})_{\gamma\in\Gamma, i\in \mathbb{N}}$ is clearly a clopen refinement of $\mathcal{A}$. Finally it is star countable: to see this fix $i_o \in N$ and $\gamma_o \in \Gamma$ and consider the set $C_{\gamma_o} \cap A_{\lambda(\gamma_o, i_o)} = D$ say. Now $B_{\gamma_o, i_o} \subset D$. If $C_\gamma \cap A_{\lambda(\gamma,i)}$ meets D we must have $\gamma = \gamma_o$ as the C's are disjoint. Thus the only possible members of the family $\mathcal{D}$ having a non-empty intersection with D belong to the subfamily $\{C_{\gamma_o} \cap A_{\lambda(\gamma_o, i)}\}_{i\in \mathbb{N}}$, and this family is countable. Thus we have proved that $\mathcal{D}$ is a clopen star countable refinement for $\mathcal{A}$.

<u>Theorem 3.16</u> If $(X,\mathcal{T})$ satisfies $\mathcal{B}_3$, then $(X,\mathcal{T})$ is SCP.

<u>Proof</u> Let $\mathcal{A}$ be a clopen cover of X and let $\mathcal{B}$ be a star countable clopen refinement of $\mathcal{A}$. Then $\mathcal{B}$ may be written as the union of its component families, $\mathcal{B} = \bigcup_{\gamma\in\Gamma} \mathcal{B}_\gamma$ where $\mathcal{B}_\gamma = (B_{\gamma,i})_{i\in \mathbb{N}}$. For each γ in Γ let $C_\gamma = \bigcup_{i=1}^{\infty} B_{\gamma,i}$. The sets C_γ are open and disjoint, thus they are also closed. For fixed γ in Γ consider $C_\gamma \subset X$ as a subspace. By the lemma above we know that the countable clopen cover $(B_{\gamma,i})_{i\in \mathbb{N}}$ of C_γ has a (countable) star finite clopen refinement which

family we will call $\mathcal{D}_\gamma$. The family $\mathcal{D} = \bigcup_{\gamma \in \Gamma} \mathcal{D}_\gamma$ is a star finite clopen refinement for $\mathcal{A}$. Thus X is SCP .

<u>Theorem 3.17</u> If $(X,\mathcal{T})$ is SCP then it is CP .

<u>Proof</u> This follows from the easy observation that a star finite clopen cover is locally finite.

<u>Theorem 3.18</u> Let $(X,\mathcal{T})$ be CP and have ind X = 0 . Then Ind X = 0 .

<u>Proof</u> Let A and B be closed disjoint nonempty subsets of X . For each x in A there is a clopen set 0_x^A such that $x \in 0_x^A \subset X \setminus B$. For each x in B there is a clopen set 0_x^B such that $x \in 0_x^B \subset X \setminus A$. Finally for x in $X \setminus (A \cup B)$ there is a clopen set 0_x with $x \in 0_x \subset X \setminus (A \cup B)$. The family

$$\mathcal{O} = \{0_x^A : x \in A\} \cup \{0_x^B : x \in B\} \cup \{0_x : x \text{ is not in } A \cup B\}$$

is a clopen cover of X . There is a clopen locally finite refinement $\mathcal{B}$ for $\mathcal{O}$ as X is CP . Let

$$\mathcal{B}^A = \{0 \in \mathcal{O} : 0 \subset 0_x^A \text{ for some } x \text{ in } A\} .$$

Then $\mathcal{B}^A$ is locally finite and covers A . Because $\mathcal{B}^A$ is locally finite, $\cup \mathcal{B}^A$ is clopen and $A \subset \mathcal{B}^A \subset X \setminus B$.

Therefore X has large inductive dimension zero.

Theorem 3.19 Let $(X,\mathcal{T})$ be paracompact and have $\operatorname{Ind} X = 0$. Then X satisfies property $\mathcal{B}_1$, that is every open cover has a clopen locally finite refinement.

Proof Let $\mathcal{U}$ be an open cover of X and let $\mathcal{B} = (B_\gamma)_{\gamma \in \Gamma}$ be a locally finite open refinement of $\mathcal{U}$. Because X is normal and $\mathcal{B}$ is point finite there exists an open cover $\mathcal{O} = (O_\gamma)_{\gamma \in \Gamma}$ of X such that $cl(O_\gamma) \subset B_\gamma$ for each γ in Γ. (See [35].) For each γ choose a clopen set C_γ with the property $O_\gamma \subset C_\gamma \subset B_\gamma$. Then $(C_\gamma)_{\gamma \in \Gamma}$ is a clopen locally finite refinement for $\mathcal{U}$.

Theorem 3.20 Let $(X,\mathcal{T})$ be a topological space. Then X satisfies $\mathcal{B}_1$ if and only if it satisfies $\mathcal{B}_2$. More generally, let $\mathfrak{p}$ be some property of covers of X (e.g. closed, open, clopen, etc.). Then every $\mathfrak{p}$ cover of X has a clopen σ-locally finite refinement, if and only if every $\mathfrak{p}$ cover has a clopen locally finite refinement.

Proof We need only prove that every clopen σ-locally finite cover of X has a clopen locally finite refinement. Let $\mathcal{E} = (E_{\alpha,n})_{\alpha \in A,\, n \in \mathbb{N}}$ be a clopen cover of X with the property that for fixed n_o, $(E_{\alpha,n_o})_{\alpha\in A}$ is a locally finite family. For each n in $\mathbb{N}$ let $W_n = \bigcup_{\alpha\in A} E_{\alpha,n}$ then $(W_n)_{n\in\mathbb{N}}$ is a clopen cover of X. By lemma 3.14 above there exists a locally finite clopen refinement $(G_n)_{n\in\mathbb{N}}$ for $(W_n)_{n\in\mathbb{N}}$ with the property $G_n \subset W_n$ for each n in $\mathbb{N}$. Now the family of subsets $\mathcal{H} = (G_n \cap E_{\alpha,n})_{\alpha\in A, n\in\mathbb{N}}$ refines $\mathcal{E}$, is a cover for X and is clopen. Each x in X has a neighborhood meeting only a finite number of the G_n's and for each such n the point x has a neighborhood meeting at most finitely many $E_{\alpha,n}$. This completes the proof.

<u>Theorem 3.21</u> Let $(X,\mathcal{T})$ satisfy $\mathcal{B}_1$. Then X is CP.

<u>Proof</u> Immediate.

<u>Notes</u>:

(i) P. Roy's example [56] is one of a paracompact (even metrizable) space having small inductive dimension zero but not being CP.

(ii) The condition CP is weaker than any of the sufficient conditions mentioned at the beginning of this section at least for paracompact spaces.

From Theorems 3.21, 3.19 and 3.18 we obtain the main result of this section:

<u>Theorem 3.22</u> Let $(X,\mathcal{T})$ be a paracompact topological space. Then X has large inductive dimension zero if and only if it has small inductive dimension zero and is clopen-paracompact.

From Theorem 3.22 it follows that, at least for paracompact spaces, clopen paracompactness is the link between the two notions small and large inductive dimension zero. Examples of CP spaces abound:

(i) Any discrete or indiscrete space is CP .

(ii) Any connected space or disconnected space with at most a countable number of components (in particular the rational numbers) is CP .

(iii) Any paracompact space having large inductive dimension zero (and hence the irrational numbers) is CP .

(iv) Every Lindelof space is CP as is every strongly paracompact space.

(v) Every locally compact paracompact space is strongly paracompact and thus CP . Hence every paracompact manifold is CP .

Example (i) shows that CP spaces do not, in general, satisfy any of the separation axioms. P. Roy's space Δ given in [56] and [57] is an example of a paracompact space which is not CP . If X is an infinite set, p a fixed element of X and

$$\mathcal{T} = \{A \subset X : p \in A\} \cup \{\phi\}$$

then $(X,\mathcal{T})$ is CP (connected) and not paracompact. These examples show that there are no implications between CP and paracompactness without additional hypotheses being made. However we will show that CP spaces may be characterized by the existence of families of real valued functions in a manner similar to the way paracompact spaces have been characterized by E. Michael in [40].

<u>Definition 3.23</u> Let $(X,\mathcal{T})$ be a topological space and $H = \{0\} \cup \{\frac{1}{n} \mid n = 1, 2, \ldots\}$. Let H have the usual subspace topology. Let $\mathfrak{F} = (f_\lambda)_{\lambda \in \Lambda}$ be a family of continuous functions, $f_\lambda : X \to H$ for each λ in Λ. We say $\mathfrak{F}$ is a <u>harmonic partition of unity</u> (HPU) if

$$\sum_{\lambda \in \Lambda} f_\lambda(x) = 1$$

for all x in X .

Now let $\mathcal{C} = (C_\alpha)_{\alpha \in A}$ be a cover of X and let $\mathfrak{F}$ be a HPU on X . We say that $\mathfrak{F}$ <u>is subordinate to</u> $\mathcal{C}$

if the family

$$\mathcal{B} = (f_\lambda^{-1}(H \setminus \{0\}))_{\lambda\in\Lambda}$$

is a refinement of $\mathcal{U}$. We say that an HPU $\mathcal{F}$ is <u>locally finite</u> if the family of sets $\mathcal{B}$ is a locally finite family.

<u>Lemma 3.24</u> For each j in $\mathbb{N}$, the natural numbers, there is an $\epsilon_j > 0$ such that for all $\{m_1, \ldots, m_k : k \leq j\} \subset \mathbb{N}$, if $\sum_{i=1}^{k} \frac{1}{m_i} < 1$ then $\sum_{i=1}^{k} \frac{1}{m_i} \leq 1 - \epsilon_j$.

<u>Proof</u> The theorem clearly is true for $j = 1$. Assume that it is true for $j = 1, 2, \ldots, n$ and suppose that it is false for $j = n + 1$. That is for all $\epsilon > 0$ there is a subset $\{m_1(\epsilon), \ldots, m_k(\epsilon) : k \leq n + 1\} \subset \mathbb{N}$ such that

$$0 < 1 - \sum_{i=1}^{k} \frac{1}{m_i(\epsilon)} < \epsilon . \tag{1}$$

Note that necessarily $k = n + 1$ by the inductive hypothesis. Let $\epsilon > 0$ be given. By (1) if $1 \leq \ell \leq n + 1$ then

$$0 < 1 - \frac{1}{m_\ell(\epsilon)} < \epsilon + \sum_{\substack{i=1 \\ i\neq \ell}}^{n+1} \frac{1}{m_i(\epsilon)} < \epsilon + 1 + \epsilon_n$$

again by the inductive hypothesis since $\sum_{\substack{i=1 \\ i\neq \ell}}^{n+1} \frac{1}{m_i(\epsilon)} < 1$.

Therefore $\epsilon_n - \epsilon < \frac{1}{m_\ell(\epsilon)}$ for each ℓ. Choose $\epsilon = \epsilon_n/2$. Then

$$m_\ell(\epsilon_n/2) < \frac{1}{(\epsilon_n - \epsilon)} = \frac{1}{(\epsilon_n/2)} .$$

Thus the number of distinct choices of sets of integers $\{m_1(\epsilon), \ldots, m_{n+1}(\epsilon)\}$, satisfying (1) with $\epsilon = \epsilon_n/2$, is finite. For this value of ϵ let δ be the maximum value of the sums $\sum_{i=1}^{n+1} \frac{1}{m_i(\epsilon)}$. Then $0 < \delta < 1$. Now let $\overline{\epsilon} = 1 - \delta$ in (1). There is a set $\{m_1(\overline{\epsilon}), \ldots, m_{n+1}(\overline{\epsilon})\} \subset \mathbb{N}$ such that

$$0 < 1 - \sum_{i=1}^{n+1} \frac{1}{m_i(\epsilon)} < \overline{\epsilon} = 1 - \delta < \epsilon . \tag{2}$$

Thus $\delta < \sum_{i=1}^{n+1} \frac{1}{m_i(\overline{\epsilon})}$. Inequality (2) implies

$$0 < 1 - \sum_{i=1}^{n+1} \frac{1}{m_i(\overline{\epsilon})} < \epsilon$$

and thus $\delta \geq \sum_{i=1}^{n+1} \frac{1}{m_i(\overline{\epsilon})}$, a contradion. We have proved the inductive step and thus completed the proof of the lemma.

<u>Theorem 3.25</u> Let $(X, \mathcal{T})$ be a topological space. The following are equivalent:

(i) X is clopen-paracompact;

(ii) for each clopen cover $\mathcal{C}$ of X there exists a locally finite harmonic partition of unity $\mathfrak{F}$ on X such that $\mathfrak{F}$ is subordinate to $\mathcal{C}$;

(iii) for each clopen cover $\mathcal{C}$ of X there exists an harmonic partition of unity $\mathfrak{F}$ on X such that $\mathfrak{F}$ is subordinate to $\mathcal{C}$.

<u>Proof</u> (ii) ⇒ (iii) is immediate. We will show that (i) ⇔ (ii) and that (iii) ⇒ (i). We have included the proof of (ii) ⇒ (i) for the sake of interest.

(i) ⇒ (ii) Let $\mathcal{C} = (C_\gamma)_{\gamma \in \Gamma}$ be a clopen cover of X and let $\mathcal{B} = (B_\lambda)_{\lambda \in \Lambda}$ be a clopen locally finite refinement for $\mathcal{C}$. For each λ in Λ let

$$f_\lambda = \chi_{B_\lambda} ,$$

the characteristic function of B_λ . Then, as $\mathcal{B}$ is locally finite, the function $f(x) = \sum_{\lambda \in \Lambda} f_\lambda(x)$ is continuous. Let $\varkappa_\lambda(x) = \dfrac{f_\lambda(x)}{f(x)}$ for each λ in Λ and x in X . Then $(\varkappa_\lambda)_{\lambda \in \Lambda}$ is a HPU subordinate to $\mathcal{C}$. Indeed $\varkappa_\lambda^{-1}(H \setminus \{0\}) = B_\lambda$ for each λ in Λ .

(ii) ⇒ (i) Let $\mathcal{C}$ be a clopen cover of X and $\mathfrak{F} = (f_\lambda)_{\lambda \in \Lambda}$ a subordinate locally finite HPU . Then the family of sets $\mathcal{B} = (f_\lambda^{-1}(H \setminus \{0\}))_{\lambda \in \Lambda}$ is an open locally finite refinement for $\mathcal{C}$. We will show that each of the members of $\mathcal{B}$ is a closed set. To this end, for fixed λ in Λ , let $f_\lambda^{-1}(\{0\}) = A_\lambda$. If $A_\lambda \neq \phi$ let $y \in A_\lambda$. There exists a neighborhood V of y on which only a finite number of the functions $(f_\lambda)_{\lambda \in \Lambda}$ are non zero. Let these be denoted $\{f_{\lambda_1}, \ldots, f_{\lambda_n}\}$. Then, for all x in V , we have $\sum_{i=1}^{n} f_{\lambda_i}(x) = 1$. If λ is not in $\{\lambda_1, \ldots, \lambda_n\}$ then $y \in V \subset A_\lambda$. If not suppose that $\lambda = \lambda_n$. For each ε with $0 < \varepsilon < 1$ let

$$N_\epsilon = V \cap f_\lambda^{-1}([0, \varepsilon) \cap H) .$$

Then N_ϵ is a neighborhood of y and $y \in N_\epsilon \subset V$. Suppose that for each ϵ in the given range there exists an x in N_ϵ such that $f_\lambda(x) \neq 0$. Then we would have at this point

$$0 < 1 - \sum_{i=1}^{n-1} f_{\lambda_i}(x) = f_{\lambda_n}(x) < \varepsilon .$$

But, by the above lemma, this is impossible. Thus there exists an $\varepsilon > 0$ such that $f_\lambda(x) = 0$ for all x in N_ϵ . Then $y \in N_\epsilon \subset A_\lambda$. Thus A_λ is an open set and hence

$f_\lambda^{-1}(H \setminus 0)$ is closed. This completes the proof that $\mathcal{B}$ is a clopen locally finite refinement for $\mathcal{C}$.

(iii) ⇒ (i) Let $\mathcal{C}$ be a clopen cover of X and let $\mathfrak{F} = (f_\lambda)_{\lambda\in\Lambda}$ be an HPU subordinate to $\mathcal{C}$. We proceed as in [40] or the equivalent [22, page 208, Lemma 2].

Let $\varphi\colon X \to H$ be any continuous function with $\varphi(x) > 0$. There exists a set $\Gamma = \{\lambda_1,\ldots,\lambda_n\} \subset \Lambda$ such that

$$1 - \sum_{i=1}^{n} f_{\lambda_i}(x) < \varphi(x)\,.$$

If we let

$$V = \{y \in X : 1 - \sum_{i=1}^{n} f_{\lambda_i}(y) < \varphi(y)\}$$

then V is a neighborhood of x and $f_\lambda(y) < \varphi(y)$ for y in V and λ in $\Lambda \setminus \Gamma$. Now let

$$f(x) = \operatorname*{lub}_{\lambda\in\Lambda} f_\lambda(x)\,.$$

Then $f\colon X \to H \setminus \{0\}$ is a continuous function. For each λ in Λ let $V_\lambda = \{x : f_\lambda(x) > \frac{1}{2} f(x)\}$. Then $\mathcal{V} = (V_\lambda)_{\lambda\in\Lambda}$ is an open covering of X refining $\mathcal{C}$. By setting $\varphi = \frac{1}{2}f$ in the above we see that $\mathcal{V}$ is also locally finite (all as in [40]). We will now show that, under the given hypotheses, each V_λ is a closed subset of X.

Let $x \in cl(V_\gamma) \setminus V_\gamma$. Then $f_\gamma(x) = \frac{1}{2}f(x) \neq 0$. But then there is a neighborhood N of x such that $f_\gamma(y) = \frac{1}{2}f(y)$ for all y in N , namely

$$N = f_\gamma^{-1}(\{f_\gamma(x)\}) \cap (\tfrac{1}{2} f)^{-1}(\{f_\gamma(x)\}) .$$

This means $N \cap V_\gamma = \phi$, a contradiction. Hence V_γ is a closed set. Therefore $\mathcal{V}$ is a clopen locally finite refinement for $\mathcal{C}$. This completes the proof of the theorem.

We return to the study of S-metric spaces as topological spaces, started at the beginning of this chapter.

<u>Theorem 3.26</u> Let $S \subset [0,\infty)$ have $\text{Ind } S = 0$ and $0 \in S$. Let $(X,\mathcal{T})$ admit a compatible S-metric d and let S admit a compatible H-metric ρ . Then the function $f: X \times X \to H$ defined by

$$f(x,y) = \rho(0, d(x,y))$$

is a continuous symmetric on X and generates the topology $\mathcal{T}$. It generates a uniform structure which is coarser than that generated by d .

<u>Proof</u> The last statement follows from the following argument: given an $\epsilon > 0$ there is a $\delta > 0$ such that if

$o \leq \ell < \delta$ and $\ell \in S$ then $\rho(0,\ell) < \varepsilon$. Thus if $d(x,y) < \delta$ then $f(x,y) < \varepsilon$. To verify the first statement we will show that $F \subset X$ is closed if and only if glb $\{f(x,y): y \in F\} = f(x,y) > 0$ for all x not in F . Let F be closed, x not be in F and $f(x,F) = 0$. Then there is a sequence $\{x_n\} \subset F$ with $f(x, x_n) < \frac{1}{n}$. Thus $\rho(0, d(x, x_n)) < \frac{1}{n}$ and $d(x, x_n) \to 0$ which implies $x_n \to x$. Thus we must have $f(x,F) > 0$. Now suppose that $f(x,F) > 0$ for all x not in F . If x is not in F let $\varepsilon = f(x,F)$ and let $Q = \{y: f(x,y) < \varepsilon\}$. Then Q is open, $Q \cap F = \phi$ and $x \in Q$. Thus F is closed. These observations show that f generates the topology $\mathcal{T}$.

Theorem 3.27 Let S be as in Theorem 3.26 above and let the topological space $(X,\mathcal{T})$ be S-metrizable. Then there exists a family $\{\rho_x\}_{x\in X}$ of continuous H-pseudometrics on X generating the topology $\mathcal{T}$ in the sense that $\mathcal{T}$ is the weakest topology on X for which each of the pseudometrics is continuous.

Proof Let $d: X \times X \to S$ be a compatible S-metric on X and let, for each x_o in X ,

$$\rho_{x_o}(x,y) = \rho(d(x, x_o), d(x_o, y)) ,$$

where x and y are in X and ρ is a compatible H-metric on S. Then each pseudometric ρ_{x_o} is continuous.

Let $B(x_o, \varepsilon) = \{z \in X : d(x_o, z) < \varepsilon\}$. Because ρ generates the subspace topology on S there is a $\delta > 0$ such that if $\rho(0, \ell) < \delta$ then $|0 - \ell| < \varepsilon$. Then if $\rho_{x_o}(x_o, y) < \delta$ we have $\rho(0, d(x_o, y)) < \delta$ which means $d(x_o, y) < \varepsilon$ or $y \in B(x_o, \varepsilon)$. This shows that the family of continuous pseudometrics $\{\rho_x\}_{x \in X}$ generates the topology $\mathcal{T}$.

<u>Remark</u> The family $\{\rho_x\}$ generates a uniform structure which is not in general comparable with the uniform structure generated by the original metric d.

Let $b_n \downarrow 0$ be a sequence of distinct points and let $S = [0,1] \setminus \{b_n\}$. Below we will obtain an analogue to Theorem 3.27 for S-metrizable topological spaces.

Let $(X, \mathcal{T})$ be an S-metrizable topological space and let $d: X^2 \to S$ be a metric generating the topology $\mathcal{T}$. Define a function a_1 on X^2 as in Theorem 2.23 above. The function a_1 has values in X_1. For each $m \in \mathbb{N}$ let $a_m(x,y) = \text{glb}\ \{a_1(x, x_1) + a_1(x_1, x_2) + \ldots + a_1(x_{m-1}, y)\}$

where the glb is taken over all (m-1) - tuples $(x_1, \ldots, x_{m-1})$ of elements of X . For each m in $\mathbb{N}$ let

$$X_m = \{ \sum_{i=1}^{m} 2^{-n_i} \mid n_i \in \mathbb{N} , 1 \leq i \leq m\} \cup \{0\} .$$

Then a_1 has values in X_1 , and for each m , with $m > 1$, a_m has values in X_m . Furthermore

$$a_1(x,y) \geq a_2(x,y) \geq \ldots \geq a_m(x,y) \geq \ldots \geq 0 .$$

Regard X_m as a subspace of R . As such it is compact and has large inductive dimension zero. Let

$$X = \prod_{m=1}^{\infty} X_m$$

and let X have the product topology. Then X is compact, metrizable, and has Ind X = 0 (Morita [41]). Let ρ be an H-pseudometric on X compatible with its topology. For each $x_o \in X$ define a function $\beta_{x_o} : X \times X \to H$ by the rule

$$\beta_{x_o}(x,y) = \rho((a_m(x, x_o)) , (a_m(x_o, y))) .$$

Then β_{x_o} is an H-pseudometric on X . Let

$$B(x, \epsilon) = \{y: \beta_x(x,y) < \epsilon\}$$

be defined for each x in X and $\epsilon > 0$. Then

Theorem 3.28 The family of H-pseudometrics $\{\beta_x\}_{x\in X}$ generates the metric topology on X in that the family $(B(x,\epsilon))_{\epsilon>0}$ is a fundamental system of neighborhoods of x for each x in X.

Proof For each m in $\mathbb{N}$ the function a_m is continuous at every point in Δ. This follows from the inequality $0 \leq a_m(x,y) \leq a_1(x,y)$ and the continuity of the function a_1. Also $a_m(x,x) = 0$ for each x in X. Let $x_n \to x_o$ in X. Then

$$\lim_{n\to\infty} \beta_{x_o}(x_o, x_n) = \lim_{n\to\infty} \rho((a_m(x_o, x_o)), (a_m(x_o,x_n))),$$

$$= \rho((0), (0)) = 0 ,$$

because ρ is continuous and X has the product topology. This means $\beta_{x_o}(x_o, x)$ is a continuous function of x at $x = x_o$. Given $\epsilon > 0$ there is a neighborhood M of x_o such that if $y \in M$ then $\beta_{x_o}(x_o, y) < \epsilon$. This means $M \subset B(x_o, \epsilon)$ and $B(x_o, \epsilon)$ is a neighborhood of x. Now let P be an open set and let $x \in P$. Then there is an $\epsilon > 0$ such that $B^d(x,\epsilon) \subset P$ where

$$B^d(x,\epsilon) = \{y : d(x,y) < \epsilon\} .$$

There is a $\delta_1 > 0$ such that if $a_1(x,y) < \delta_1$ then

$d(x,y) < \varepsilon$. The metric

$$\theta((b_m), (c_m)) = \sum_{m=1}^{\infty} |b_m - c_m| 2^{-m}$$

generates the same topology as ρ on X . Thus there is a $\delta_2 > 0$ such that if $\rho((0), (a_m(x,y))) < \delta_2$ then $\theta((0), (a_m(x,y))) < \frac{1}{2}\delta_1$ This implies $a_1(x,y) < \delta_1$ and thus $d(x,y) < \varepsilon$. Hence $B(x, \delta_2) \subset P$. This concludes the proof that the family $\{B(x,\varepsilon): \varepsilon > 0, x \in X\}$ is a fundamental system of neighborhoods for the metric topology of X .

<u>Remark</u> The functions a_m being the greatest lower bounds of families of continuous functions are upper semi-continuous on $X \times X$. When they are continuous the sets $B(x,\varepsilon)$ are open and the family $\{\beta_x\}_{x \in X}$ is a set of continuous pseudometrics on X which generates the metric topology.

<u>Topologies generated by single functions on X^2</u>

Let X be a set and let $d: X^2 \to R$ be a function. The following is a list of possible axioms for d :

1. $d(x,y) \geq 0$ for all x and y and $d(x,x) = 0$ for all x ,

2. If $d(x,y) = 0$ then $x = y$,

3. $d(x,y) = d(y,x)$ for all x and y ,

4. $d(x,y) \leq d(x,z) + d(z,y)$ for all x, y and z ,

5. $d(x,y) \leq \max \{d(x,z), d(z,y)\}$ for all x, y and z .

<u>Definition 3.29</u> We will consider the following combinations of axioms:

$\mathcal{P}_1 = \{1,2,3,5\}$ ultrametric , $\mathcal{P}_{10} = \{1,3,5\}$ pseudo-ultrametric,

$\mathcal{P}_2 = \{1,2,3,4\}$ metric, $\mathcal{P}_{20} = \{1,3,4\}$ pseudometric,

$\mathcal{P}_3 = \{1,2,3\}$ semimetric, $\mathcal{P}_{30} = \{1,3\}$,

$\mathcal{P}_4 = \{1,2\}$, $\mathcal{P}_{40} = \{1\}$,

$\mathcal{P}_5 = \{1,2,4\}$ quasimetric, $\mathcal{P}_{50} = \{1,4\}$ pseudoquasi-metric.

Define a partial order on the set of properties by setting $\mathcal{P}_\alpha > \mathcal{P}_\beta$ if for all d , if d satisfies $\mathcal{P}_\alpha$, then d satisfies $\mathcal{P}_\beta$, (see Figure 6 below).

A rule for the generation of topologies is defined as follows:

<u>Definition 3.30</u> Let X be a set and d a function on X^2 satisfying $\mathcal{P}_\lambda$ for some λ in

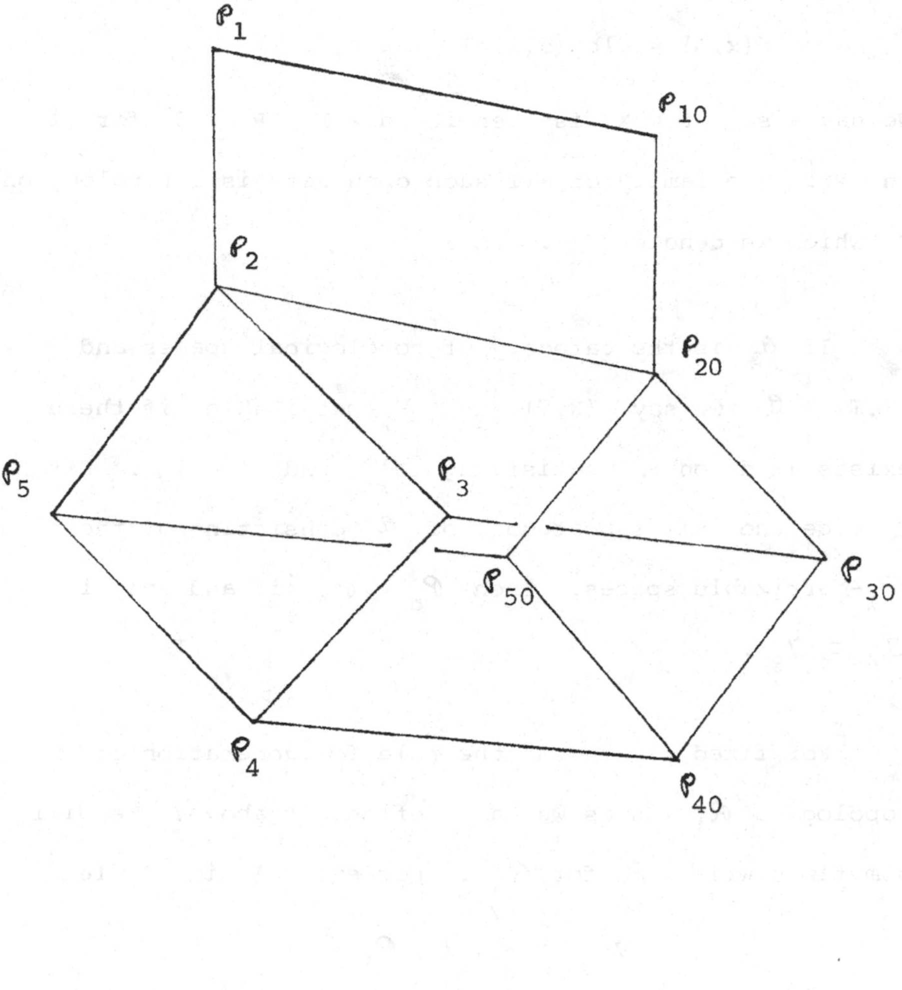

Figure 6

Relationships between properties defining common types of metrics

$\Lambda = \{i: 1 \leq i \leq 5\} \cup \{(i,0): 1 \leq i \leq 5\}$. If $A \subset X$ is a non empty subset and x belongs to X let

$$d(x,A) = \text{glb } \{d(x,a): a \in A\} .$$

We say a set $P \subset X$ is open if $d(x,X - P) > 0$ for all x in P . The family of all such open sets is a topology on X which we denote $\mathcal{T}_d$.

If $\mathcal{A}$ is the category of topological spaces and $(X,\mathcal{T}) \in \mathcal{A}$ we say $(X,\mathcal{T})$ is $\mathcal{P}_\lambda$-metrizable if there exists a d on X satisfying $\mathcal{P}_\lambda$ and $\mathcal{T} = \mathcal{T}_d$. Let $\mathcal{T}_\lambda$ be the full subcategory of $\mathcal{A}$ consisting of the $\mathcal{P}_\lambda$-metrizable spaces. Then $\mathcal{P}_\alpha > \mathcal{P}_\beta$ if and only if $\mathcal{T}_\alpha \subset \mathcal{T}_\beta$.

For fixed λ in Λ the rule for generation of topologies $\mathcal{R}_\lambda$ is as we have defined it above. We will sometimes write $\mathcal{R}$ for $\mathcal{R}_\lambda$. For each λ in Λ let

$$V_\lambda = (\mathcal{T}_\lambda, \mathcal{P}_\lambda, \mathcal{R}_\lambda) .$$

Remark In the quasimetric case we have given a rule for the generation of the "right" topology. There is a corresponding theory when the rule corresponds to the generation of the "left" topology and the two partially ordered sets $P(V,\mathcal{S})$ so formed are isomorphic.

Theorem 3.31 Let $[S] = [S, V_1, \mathcal{H}]$ for each $S \in \mathcal{H}$. Then $[0] \prec [D] \prec [H]$ is a complete presentation for $P(V_1, \mathcal{H})$.

Proof This follows from the theorem of J. de Groot [27] (if the topological space $(X, \mathcal{T})$ has a compatible metric satisfying ρ_1 then $\text{Ind } X = 0$) and Broughan [13, Theorem 1].

Theorem 3.32 $P(V_i, \mathcal{H}) = P(V_{io}, \mathcal{H})$ for $1 \leq i \leq 5$.

Proof Let i be fixed with $1 \leq i \leq 5$ and let $[S, V_\lambda] = [S, V_\lambda, \mathcal{H}]$ for $S \in \mathcal{H}$ and $\lambda \in \Lambda$. Because $\mathcal{T}_{io} \supset \mathcal{T}_i$, if S and T in $\mathcal{H}$ are such that $[S, V_i] \neq [T, V_i]$ then $[S, V_{io}] \neq [T, V_{io}]$. Let $[S, V_i] \prec [T, V_i]$ and let $(X, \mathcal{T}) \in \mathcal{T}_{io}$. Let d on X^2 satisfy $\mathcal{P}_{io}$ and $d(X^2) \subset S$. Define an equivalence relation $\sim$ on X by setting $x \sim y$ if and only if $d(x,y) = 0$, let Y be the corresponding quotient space and $q\colon X \to Y$ the quotient map. Then

$$h = d \circ (q \times q)$$

is a metric on Y compatible with the quotient topology and $Y \in \mathcal{T}_i$. By hypothesis there exists a function, ℓ say, on Y^2 satisfying $\mathcal{P}_i$, generating the quotient topology, and taking values in T . Define a function satisfying $\mathcal{P}_i$ on X^2 by

$$m = \ell \circ (q \times q) .$$

Then $\mathcal{T} = \mathcal{T}_m$ and $m(X^2) \subset T$. This shows that $[S,V_{io}]$ is less than $[T,V_{io}]$ and it follows that $[S,V_i] = [T,V_i]$ if and only if $[S,V_{io}] = [T,V_{io}]$. In other words $P(V_i,\mathcal{H}) = P(V_{io},\mathcal{H})$.

<u>Theorem 3.33</u> Let $[S] = [S,V_3,\mathcal{H}]$ for each $S \in \mathcal{H}$. Then $[0] < [D] < [H]$ is a complete presentation for $P(V_3,\mathcal{H})$.

<u>Proof</u> This follows from the proof of J.R. Boyd's theorem [12, Theorem 1].

<u>Theorem 3.34</u> Let $[S] = [S,V_4,\mathcal{H}]$ for each $S \in \mathcal{H}$. Then $[0] < [D] < [H]$ is a complete presentation for $P(V_4,\mathcal{H})$.

<u>Proof</u> By the result of Brown [16] a space $(X,\mathcal{T})$ is in $\mathcal{J}_4$ if and only if it is Hausdorff and first countable. Let $(X,\mathcal{T})$ satisfy these properties and let $N_x = (B_{x,n})_{n\in\mathbb{N}}$ be a base for the neighborhoods of x for each x in X . We may assume that $B_{x,n+1} \subset B_{x,n}$ for

all n in $\mathbb{N}$. Let

$$d(x,y) = \inf \{1/n: y \in B_{x,n}\} .$$

Then d satisfies ρ_4 and generates the topology $\mathcal{T}$. The map d has values in H . The result of the theorem follows from these remarks.

<u>Conjecture 3.35</u> $P(V_5, \mathcal{H}) \neq P(V_2, \mathcal{H})$.

<u>Theorem 3.36</u> Let $[S] = [S, V_5, \mathcal{H}]$ for each S in $\mathcal{H}$. Then $[0] < [D] < [H] < [I]$ is a presentation for $P(V_5, \mathcal{H})$ and the first three terms form the initial segment.

<u>Proof</u> The proof, in the main, is similar to that of Theorem 3.9. We will show here that $[H] \neq [I]$: let $h(x,y) = \min \{1, |x - y|\}$ and let $R = (R, \mathcal{T}_h)$, the real numbers with the usual topology. Suppose there exists a quasimetric d on R , generating the topology $\mathcal{T}_h$, and taking its values in $H \subset [0,\infty)$. Then $(R, \mathcal{T}_d) = (R, \mathcal{T}_h)$. Let $d^t(x,y) = d(y,x)$ for each x and y in R and set $R^t = (R, \mathcal{T}_{d^t})$. Consider the map $d: R \times R^t \to H$. For all a, b, x and y in R we have

$$|d(x,y) - d(a,b)| \leq d^t(a,x) + d(b,y)$$

because d satisfies Axiom 4. It follows that d is continuous on $\mathbb{R} \times \mathbb{R}^t$. If $a \in \mathbb{R}^t$ then $\mathbb{R} \times \{a\}$ is connected and so $\{d(x,a): x \in \mathbb{R}\} = \{0\}$. Because this is true for all a in $\mathbb{R}^t$, $d(x,y) = 0$ for all x and y . This contradiction shows that a quasi-metric d with values in H does not exist. Therefore $[H] \neq [I]$.

Remark 3.37 Let $(X,\mathcal{T})$ be a topological space and $\mathcal{B}$ a base for $\mathcal{T}$. We say $\mathcal{B}$ is an σ-Q-base for $\mathcal{T}$ if $\mathcal{B} = (B(n,\lambda): n \in \mathbb{N}, \lambda \in I_n)$ and, for each n in $\mathbb{N}$ and x in X , $\cap\{B(n,\lambda): x \in B(n,\lambda), \lambda \in I_n\}$ is open, Fletcher and Lindgren [25, § 3]. A T_1 topological space $(X,\mathcal{T})$ is $[H,V_5,\aleph]$-metrizable if and only if there exists a σ-Q-base for the topology $\mathcal{T}$. This follows from the proof of the above theorem and Fletcher and Lindgren [25, Theorem 3.2].

A T_1 topological space $(X,\mathcal{T})$ is $[I,V_5,\aleph]$-metrizable if and only if, for each x in X , there is a base $U_x = \{U(x,n): n \in \mathbb{N}\}$ for the neighborhood system of x such that $X = U(x,1)$ and such that if $y \in U(x, n+1)$ then $U(y, n+1) \subset U(x,n)$, by Stoltenberg [63, Theorem 2.1].

Now let $\mathcal{Q}$ be the class of second countable T_1 topological spaces. Note that $\mathcal{Q} \subset \mathcal{J}_5$ by Norman [45,

Theorem 2] and Sion and Zelmer [60, Theorem 2.2].

Theorem 3.38 Let $V_{\mathcal{A}} = (\mathcal{A}, \mathcal{P}_5, \mathcal{R}_5)$ and, for each S in $\mathcal{H}$, $[S] = [S, V_{\mathcal{A}}, \mathcal{H}]$. Then

$$[O] < [D] < [H]$$

is a complete presentation for $P(V_{\mathcal{A}}, \mathcal{H})$.

Proof This follows from Theorem 3.36 and Fletcher and Lindgren [25, Theorem 3.3].

Remark If $\mathcal{B}$ is the class of second countable metrizable spaces and $V_{\mathcal{B}} = (\mathcal{B}, \mathcal{P}_2, \mathcal{R}_2)$ then, by Theorem 3.11 and Theorem 3.38, $P(V_{\mathcal{A}}, \mathcal{H}) \neq P(V_{\mathcal{B}}, \mathcal{H})$.

Topologies generated by families of functions

Let $\mathcal{D}$ be the class of topological spaces with the weak topology induced by sets of real valued function. If $(X, \mathcal{T}) \in \mathcal{D}$ and $F\colon X \to R$ is a family of functions we will say F satisfies $\mathcal{P}$. If $\mathcal{T}$ is the weakest topology on X such that each $f \in F$ is continuous we will write $\mathcal{T} = \mathcal{T}_F$. If Y is a set and $F\colon Y \to R$ a family of real valued functions of Y then by the rule $\mathcal{R}$ we mean that the topology on Y is the weak topology generated by the functions F . Let $\mathcal{S}$ be the family of nonempty subsets of R .

<u>Theorem 3.39</u> Let $V = (\mathcal{D}, \mathcal{P}, \mathcal{R})$ and $[S] = [S, V, \mathcal{S}]$ for each S in $\mathcal{S}$. Then

$$[0] < [D] < [I]$$

is a complete presentation for $P(V, \mathcal{S})$.

<u>Proof</u> (a) For all S in $\mathcal{S}$, $[0] < [S]$: because $S \neq \emptyset$ there exists an element x in S . If $(X, \mathcal{T})$ is a topological space and $F: X \to \{0\}$ satisfies $\mathcal{T} = \mathcal{T}_F$ then the topology $\mathcal{T}$ is indiscrete. Let $g: X \to \{x\}$ be the constant function. Then $\mathcal{T} = \mathcal{T}_{\{g\}}$. Thus $0 < \{x\} < S$ and so $[0] < [S]$.

(b) For all S in $\mathcal{S}$, $[S] < [I]$: we need only prove that $[R] < [I]$. Let $(X, \mathcal{T})$ be a space and $F: X \to R$ a family of functions with $\mathcal{T} = \mathcal{T}_F$. Let, for all x in R

$$g(x) = \tfrac{1}{2} \arctan (x) + \tfrac{1}{2} .$$

Then the family $G = \{g \circ f: f \in F\}$ satisfies $\mathcal{T}_G = \mathcal{T}_F$ and $G: X \to I$. Therefore $R < I$ and so $[R] < [I]$.

(c) If $[0] < [S]$ and $[0] \neq [S]$ than $[D] < [S]$: because $[0] \neq [S]$ there exists $\{a,b\} \subset S$ with $a \neq b$.

Let $(X,\mathcal{T})$ be a space and $F: X \to D$ a family of functions with $\mathcal{T} = \mathcal{T}_F$. Let $G = \{h \circ f: f \in F\}$ where $h(1) = a$ and $h(0) = b$. Then $\mathcal{T}_F = \mathcal{T}_G$. The conclusion follows from this.

(d) If $S \in \mathcal{J}$ is such that the interior of S in the usual topology of $\mathbb{R}$ is not empty then $[I] < [S]$: let $k: I \to S$ be an embedding. If $(X,\mathcal{T})$ is such that there exists $F: X \to I$ with $\mathcal{T} = \mathcal{T}_F$ let $G = \{k \circ f: f \in F\}$. This implies that $[I] < [S]$.

(e) If $S \in \mathcal{J}$ is such that the interior of S is empty then $[S] < [D]$: there exists an embedding $q: S \to D^{\aleph_0}$ where $D^{\aleph_0}$ is the countable product of copies of D with the product topology. Let $g_n: D^{\aleph_0} \to D$, $n \in \mathbb{N}$, be the projections onto the factors. If $(X,\mathcal{T})$ is such that there exists $F: X \to S$ with $\mathcal{T} = \mathcal{T}_F$ let $G = \{g_n \circ q \circ f: f \in F , n \in \mathbb{N}\}$. Then $\mathcal{T}_G = \mathcal{T}_F$ and consequently $[S] < [D]$.

(f) Simple examples show that $[0] \neq [D]$ and $[D] \neq [I]$. The proof follows from these assertions.

Chapter IV : Algebraic Structures

In this chapter we will widen the scope of the theory developed in the previous chapters. Let $\mathcal{A}$ be a real-generated category with datum V and let $\mathcal{S} \in \mathcal{J}$. The degree of complexity of the partially ordered sets $P(V, \mathcal{S})$ is a measure of the dependence of $\mathcal{A}$ upon the structure of the real numbers and of the degree of complexity of $\mathcal{A}$ itself. This method of investigation and the notion of invariants $[S, V, \mathcal{S}]$ will be applied to algebraic categories where the structure of objects is partially generated by a single function with values in subsets S of $\mathbb{R}$. The range category "subsets of $\mathbb{R}$ " may be exchanged for some other category although we will not develop this idea.

Properties analogous to "[S]-metrizable" arise and we interpret the properties in a manner intrinsic to the category under consideration.

For the most part we are interested in topological structures. The investigation will be limited also to the most common type of function and algebraic structure which has been studied traditionally. Topological groups, rings, fields, and vector spaces are considered.

Metrizable Topological Abelian Groups

Let $\mathcal{G}$ be the category of metrizable topological abelian groups (written additively) where the morphisms are continuous group homomorphisms. If $(G,\mathcal{T})$ is in $\mathcal{G}$, then there exists a translation invariant metric d on G generating the topology $\mathcal{T}$, Bourbaki [11, page 161, Proposition 2]. The metrics on groups will be chosen to satisfy the metric axioms 1, 2, 3 and 4 of Chapter III, Definition 3.29 above and 6. $d(x+a, y+a) = d(x,y)$ for all x, y, and a in G.

Definition 4.1 Let $\mathcal{P}_6 = \{1,2,3,4,6\}$ and $V_6 = (\mathcal{G}, \mathcal{P}_6, \mathcal{R})$ in keeping with the notation used previously. The rule $\mathcal{R}$ is the usual rule for the generation of a metric topology.

Theorem 4.2 Let $[S] = [S,V_6,\mathcal{H}]$ for each S in $\mathcal{H}$. Then

$$[0] < [D] < [H] < [Q^+] < [I]$$

is a presentation for $P(V_6,\mathcal{H})$. The first four classes form the initial segment for $P(V_6,\mathcal{H})$ and $[I]$ is the maximum class.

<u>Proof</u> We will prove the theorem in stages:

(a) If $[0] \neq [S]$ then $[D] < [S]$: If $[0] \neq [S]$ then $\{0\} \neq S$ and there exists a $b > 0$ such that $\{0,b\} \subset S$. If $(G,\mathcal{T})$ is a topological group with compatible translation invariant metric d taking values in D then $h = b.d$ is translation invariant, has values in S, and $\mathcal{T}_h = \mathcal{T}$. Thus $[D] < [S]$.

(b) If $[D] < [S]$ and $[D] \neq [S]$ then $[H] < [S]$: If 0 is not in the set $cl(S - \{0\})$ and $(G,\mathcal{T})$ is in $\mathcal{G}$ then the topology $\mathcal{T}$ is discrete and so the metric $d(x,y) = 1$ for all $x \neq y$ in G generates the topology and is translation invariant. This shows that $[S] < [D]$ and so we must have $0 \in cl(S - \{0\})$. Suppose that $cl(S)$ is not a neighborhood of zero in R^+ and let $(G,\mathcal{T})$ be a topological group with a compatible translation invariant metric d having values in S. Regard G as a uniform space. Because d is translation invariant it generates the uniform structure of G. By Theorem 2.4 there exists a metric h on G, uniformly equivalent to d and having its values in H. Let

$$\ell(x,y) = \max_{a \in G} h(x+a,\ y+a)\ .$$

Then ℓ is a metric on G, is translation invariant, and, as the following argument will show, generates the same topology as h and hence the same topology as d :

Let $(x_n)_{n \in \mathbb{N}}$ be a sequence in G and x a point in G satisfying $\ell(x_n, x) \to 0$. Then $h(x_n + 0, x + 0) \to 0$ and thus $x_n \to x$ in $\mathcal{T}$. Because d and h are uniformly equivalent, given $\epsilon > 0$ there exists a $\delta > 0$ such that for all x and y in G , if $d(x,y) < \delta$ then $h(x,y) < \epsilon$. There exists an $\eta > 0$ such that, for all x and y in G , if $h(x,y) < \eta$ then $d(x,y) < \delta$. Therefore, if $h(x,y) < \eta$ then $d(x+a, y+a) < \delta$ because d is translation invariant. Therefore if x and y in G are such that $h(x,y) < \eta$ then $h(x+a, y+a) < \epsilon$ for all a in G . If $h(x_n, x) \to 0$, given $\epsilon > 0$ there exists an $\eta > 0$ and an N in $\mathbb{N}$ (η depends on ϵ and N depends on η) such that if $n \geq N$ then $h(x_n, x) < \eta$. But then, by what we have just proved, $h(x_n+a, x+a) < \epsilon$ for all a in G and $n \geq N$. Therefore $\ell(x_n, x) \leq \epsilon$ for all $n \geq N$. This proves that ℓ generates the same topology as d .

There exists a sequence of distinct points $(x_n)_{n \in \mathbb{N}}$ in S . What we have just proved implies that $[H] = [\{x_n : n \in \mathbb{N}\}]$ and therefore $[H] < [S]$.

(c) If $[H] < [S]$ and $[H] \neq [S]$ then $[\mathbb{Q}^+] < [S]$: If the hypotheses are satisfied, $cl(S)$, by (b) above, is a neighborhood of zero. Therefore, by Theorem 1.30, $\hom(\mathbb{Q}^+, S) \neq \emptyset$. Let $f \in \hom(\mathbb{Q}^+, S)$ and let $(G,\mathcal{T})$ be a metrizable topological group with a translation invariant metric d on G , generating $\mathcal{T}$, and taking values in $\mathbb{Q}^+$. Then $f \circ d$ is a comptable, translation invariant S-metric on G . We have shown that $[\mathbb{Q}^+] < [S]$.

(d) If S is in $\mathcal{H}$ and $[S] \neq [0]$, $[D]$, $[H]$, or $[\mathbb{Q}^+]$ then $[\mathbb{Q}^+] < [S] < [I]$: We need only prove that, for all S in $\mathcal{H}$, $[S] < [I]$. Let $(G,\mathcal{T})$ be in $\mathcal{G}$ and let d be a translation invariant metric on G satisfying $\mathcal{T} = \mathcal{T}_d$. Then $h(x,y) = \max\{1, d(x,y)\}$ is translation invariant, has values in I , and satisfies $\mathcal{T} = \mathcal{T}_h$.

(e) $[0] \neq [D]$: The group $\mathbb{Z}$ with the metric $d(n,m) = 1$ when $n \neq m$ is metrizable with a translation invariant D-metric but not [0]-metrizable.

(f) $[D] \neq [H]$: The group $\mathbb{Z}^{\mathbb{N}}$ with the translation invariant metric

$$d((x_i), (y_i)) = \{1/n: x_i = y_i \text{ for } 1 \leq i \leq n-1 \text{ and } x_n \neq y_n\}$$

is H-metrizable but not D-metrizable.

(g) $[H] \neq [\mathbb{Q}^+]$: The usual metric on the group $\mathbb{Q}$ is translation invariant and has values in $\mathbb{Q}^+$. If there existed a translation invariant metric d on $\mathbb{Q}$, generating the usual topology and having values in H, then d would necessarily be uniformly equivalent to the usual metric. The respective completions would be homeomorphic. This is impossible because one completion is connected and the other is not.

(h) $[\mathbb{Q}^+] \neq [I]$: The group $\mathbb{R}$ does not have a compatible $\mathbb{Q}^+$-metric. This completes the proof of the theorem.

<u>Remark 4.3</u> The same presentation as given in Theorem 4.2 is obtained when the class of (not necessarily abelian) groups is considered. To prove this we need make only minor changes in the proof of Theorem 4.2.

<u>Topological Rings and Fields</u>

All rings are commutative. The topology in each case will be generated by a function f satisfying various axioms. The following is a complete list of the axioms which will be considered in this context:

1. $f(x) = 0$ if and only if $x = 0$,
2. $f(x) = f(-x)$ for all x,

3. $f(x+y) \leq f(x) + f(y)$ for all x and y,

4. $f(x+y) \leq \max \{f(x), f(y)\}$ for all x and y,

5. $f(x.y) \leq \min \{f(x), f(y)\}$ for all x and y,

6. $f(x.y) \leq f(x).f(y)$ for all x and y,

7. $f(x.y) = f(x).f(y)$ for all x and y.

Clearly, 4 implies 3, 7 implies 6, and 6 implies 5 whenver $f(x) \leq 1$ and $f(y) \leq 1$.

If R is a ring and $f: R \to R^+$ a function satisfying 1 and 2 let

$$V_n = \{x \in R : f(x) < 3^{-n}\}$$

for each n in $\mathbb{N}$. Then the family of all translates of the sets $(V_n)_{n \in \mathbb{N}}$ forms a subbase for a topology on R. If f satisfies 3 or 4 then the subbase is a base and the operation

$$R \times R \to R$$
$$(x,y) \to x + y$$

is continuous. If f satisfies 3 and 5, or 3 and 6, the operation

$$R \times R \to R$$
$$(x,y) \to x.y$$

is continuous. If R is a field, R^* the set of non zero

elements of R , and f satisfies 3 and 7 then the operation

$$R^* \to R^*$$
$$x \to x^{-1}$$

is continuous.

<u>Definition 4.4</u> These considerations lead us to take the following combinations of axioms for functions on topological rings:

$\mathcal{P}_{21} = \{1,2,3,5\}$, $\mathcal{P}_{22} = \{1,2,3,6\}$, $\mathcal{P}_{23} = \{1,2,4,5\}$ and $\mathcal{P}_{24} = \{1,2,4,6\}$. For functions on topological fields we specify: $\mathcal{P}_{31} = \{1,2,3,7\}$ and $\mathcal{P}_{32} = \{1,2,4,7\}$.

The list of function properties clearly is not exhaustive. As before we have chosen the types of funtions which have been studied traditionally.

<u>Theorem</u> [Kaplansky, 31, Theorem 3] : A topological field is $\mathcal{P}_{31}$-metrizable if and only if the set of nil-potent elements is open and, if A is any subset of F bounded away from zero then A^{-1} is bounded.

If $\alpha \in \{21,22,23,24\}$ let $\mathcal{L}_\alpha$ be the category of $\mathcal{P}_\alpha$-metrizable topological rings. If $\beta \in \{31,32\}$ let $\mathcal{F}_\beta$ be the category of $\mathcal{P}_\beta$-metrizable topological fields. The rules $\mathcal{R}_\alpha$ and $\mathcal{R}_\beta$ are the usual ones described at the beginning of this section. Let $V_\alpha = (\mathcal{L}_\alpha, \mathcal{P}_\alpha, \mathcal{R}_\alpha)$ be the datum of $\mathcal{L}_\alpha$ for each α and let $V_\beta = (\mathcal{F}_\beta, \mathcal{P}_\beta, \mathcal{R}_\beta)$ be the datum for $\mathcal{F}_\beta$ for each β.

Theorem 4.5 A topological field F is $\mathcal{P}_{32}$-metrizable if and only if

(i) The nilpotent elements form an open bounded set,

(ii) If a is nilpotent and b is nilpotent or neutral then ab is nilpotent, and

(iii) If a and b are nilpotent or neutral then a + b is nilpotent or neutral.

Proof (necessity) Let $f: F \to R^+$ be a function satisfying the properties $\mathcal{P}_{32}$ and generating the topology of F. Then the set of nilpotent elements

$$P = \{x \in F : f(x) < 1\}$$

and the set of neutral elements

$$N = \{x \in F : f(x) = 1\} .$$

From the continuity of f we obtain (i), from axiom 7 we derive (ii), and (iii) follows from axiom 4.

(sufficiency) By Kaplansky [31, Theorem 1] there exists a function f on F, generating the topology, and satisfying {1,2,3,7}. The set of nilpotent elements and the set of neutral elements have the descriptions given in the first part of this proof. It follows from (iii) that if x and y in F are such that $f(x) \leq 1$ and $f(y) \leq 1$ then $f(x+y) \leq 1$. This implies that f is bounded on the prime ring of F and therefore f necessarily satsifies axiom 4, Lang [34, page 285]. This completes the proof.

Let $E = \{1/3^n : n \in \mathbb{Z}\} \cup \{0\}$.

<u>Theorem 4.6</u> Let $[S] = [S, V_{32}, \mathcal{S}]$ for each $S \in \mathcal{S}$ where $\mathcal{S}$ is the family of sets $\{S \in \mathcal{N} : S = G \cup \{0\}$ where G is a multiplicative subgroup of $\mathbb{R}^+\}$. Then

$$[D] < [E]$$

is a complete presentation for $P(V_{32}, \mathcal{S})$.

<u>Proof</u> If S is in $\mathcal{S}$ then $D \subset S \subset \mathbb{R}^+$ and therefore $[D] < [S] < [\mathbb{R}^+]$. In particular $[E] < [\mathbb{R}^+]$. We will show that $[E] = [\mathbb{R}^+]$: Let F be a topological field which admits a function f satisfying $\mathcal{P}_{32}$ and generating the topology of F . Assume that the topology is not discrete. Then by Bourbaki [9, Proposition 3(b), page 405] there exists a function v on K^*

$$v : K^* \to \mathbb{Z}$$

satisfying $v(1) = 0$, $v(xy) = v(x) + v(y)$ and $v(x+y) \geq \min\{v(x), v(y)\}$ for all x and y in K^* and a real number a with $0 < a < 1$ satisfying

$$f(x) = a^{v(x)}$$

for all x in K^* . Define a function g on K by $g(0) = 0$ and

$$g(x) = 3^{-v(x)}$$

for x in K^* . Then g satisfies $\mathcal{P}_{32}$, has values in E and generates the same topology as f . This shows that $[E] = [R^+]$. The same type of construction can then be used to prove that if $[D] < [S]$ and $[D] \neq [S]$ then $[E] < [S]$. Finally $[D] \neq [E]$ because, for example, $\mathbb{Q}_3$, the 3-adic completion of $\mathbb{Q}$, is not discrete.

<u>Theorem 4.7</u> Let $\mathcal{S} \subset \mathcal{H}$ be defined as in Theorem 4.6 and let $[S]$ be $[S, V_{31}, \mathcal{S}]$ for each S in $\mathcal{S}$. Then

$$[D] < [E] < [\mathbb{Q}^+] < [R^+]$$

is a presentation for $P(V_{31}, \mathcal{S})$ and the first three terms form the initial segment.

<u>Proof</u> Let F be a non discrete topological field with topology generating function f satisfying $\mathcal{P}_{31}$ and

taking values in $S \in \mathcal{S}$. We will prove that necessarily $[Q^+] < [S]$: By the theorem of Ostrowski (Bourbaki [9, Theorem 2, page 410]), there exists a number t with $0 < t < 1$ and an isomorphism j of F onto an everywhere dense subfield of $\mathbb{R}$ or $\mathbb{C}$ such that

$$f(x) = |j(x)|^t$$

for each x in F . Therefore S contains $(Q^+)^t$, because $j(Q.1) = Q$, 1 being the identity of F . Let K be a topological field with function g taking values in Q and satisfying $\mathcal{P}_{31}$. Then the function defined by $x \to g(x)^t$ has values in $(Q^+)^t$, generates the given topology on K and satisfies $\mathcal{P}_{31}$. This proves that $[Q^+] < [S]$ as was claimed.

Suppose that $[Q^+] < [S] < [R^+]$ is false and let F have a compatible function f satisfying $\mathcal{P}_{31}$ taking values in S . If F is not discrete, f must satisfy $\mathcal{P}_{32}$ by the first part of this proof. By Theorem 4.6 there exists a function g on F satisfying $\mathcal{P}_{32}$ and hence $\mathcal{P}_{31}$ which generates the topology of F and takes values in E . This proves that

$$[D] < [E] < [Q^+]$$

is the initial segment of $P(V_{31}, \mathcal{S})$ provided we know that

$[E] \neq [Q^+]$. But this is easy to see because Q with the usual topology does not admit a compatible function with values in E . This completes the proof.

Let F be a topological field which admits a compatible function satisfying $\mathcal{P}_{31}$ and let

$$N = \{x \in F\colon x \text{ is not nilpotent and } x^{-1} \text{ is not nilpotent}\} .$$

Then N is a multiplicative subgroup of F^* . Let

$$V(F) = F^*/ N$$

be the abelian quotient group.

<u>Theorem 4.8</u> Let F be a topological field and let $[S] = [S, V_{31}, \mathcal{S}]$ for each S in $\mathcal{S}$. Then

(a) F is $[D]$-metrizable if and only if $V(F)$ is trivial,

(b) F is $[E]$-metrizable if and only if $V(F)$ is infinite cyclic, and

(c) If F is $[Q^+]$-metrizable then $V(F)$ is countable.

<u>Proof</u> Let S be in $\mathcal{S}$ and $f\colon F \to S$ a function on a topological field satisfying $\mathcal{P}_{31}$ and generating the topology of F . Then

$$N = \{x \in F : f(x) = 1\}$$

and

$$\hat{f} : F^*/N \to f(F^*)$$
$$\dot{x} \to f(x)$$

(where x is in F^* and $\dot{x}$ represents the equivalence class to which x belongs) is an isomorphism of (ordered) abelian groups and $V(F) = f(F^*)$. We will now prove that the given conditions are necessary:

(a) If F is [D]-metrizable then there exists such an f on F with $f(F^*) \subset \{1\}$ and so $V(F)$ is trivial.

(b) If F is [E]-metrizable there exists such an f on F with $f(F^*) \subset E$ and therefore $V(F)$ is cyclic.

(c) If F is $[Q^+]$-metrizable there exists such an f on F with $f(F^*) \subset Q$ and $V(F)$ is countable.

Conversely let F be a topological field which admits a topology generating function satisfying $\mathcal{P}_{31}$. By Kaplansky [31, Theorem 1] $V(F)$ is a totally ordered abelian group and so, by Hölder [29], it is isomorphic to an additive subgroup of R. Let g be an isomorphism. There exists a number $a > 0$ such that

$$f(x) = a^{g(x)}$$

is a function on F satisfying $\mathcal{P}_{31}$. We may now complete the proof of the converse:

(a) If $V(F)$ is trivial then $f(F) = D$ and F is D-metrizable.

(b) If $V(F)$ is infinite cyclic there exists a function h on F, satisfying $\mathcal{P}_{31}$, generating the topology, and taking values in E, by Theorem 4.7.

<u>Exercise 4.9</u> Let $\mathbb{Q}(\sqrt{2})$ be the usual field extension of $\mathbb{Q}$. Prove that $[\mathbb{Q}^{+}, V_{31}, \mathcal{S}] \neq [\mathbb{Q}(\sqrt{2})^{+}, V_{31}, \mathcal{S}]$

<u>Theorem 4.10</u> Let F be a topological field and let f be a function on F satisfying $\mathcal{P}_{31}$ and generating the topology. Let F be complete in the usual uniform structure generated by f. Then $f(F)$ is closed in $\mathbb{R}$.

<u>Proof</u> By hypothesis the additive group of F is complete. By Bourbaki [10, Proposition 8, page 283] the multiplicative group of units F^* is complete. It is also metrizable. Let $N = \{x \in F : f(x) = 1\}$. Then N is a closed subgroup of F^*, and so, by Bourbaki [11, Chapter IX, 3.1, Proposition 4], the group F^*/N is complete. Therefore $f(F^*)$ is a complete multiplicative subgroup of $\mathbb{R}^+$ and hence $f(F)$ is closed in $\mathbb{R}$.

<u>Theorem 4.11</u> Let $\mathcal{C}$ be the class of complete topological fields, let $\mathcal{S}$ be defined as in Theorem 4.6, let the rule $\mathcal{R}$ for the generation of topologies be the usual one, and let $V_{\mathcal{C}} = (\mathcal{C}, \mathcal{P}_{31}, \mathcal{R})$. Let

$[S] = [S, V_{\mathcal{a}}, \mathcal{S}]$ for each S in $\mathcal{S}$. Then

$$[D] \prec [E] \prec [R^+]$$

is a complete presentation for $P(V_{\mathcal{a}}, \mathcal{S})$.

Proof The map

$$[S, V_{31}, \mathcal{S}] \to [S, V_{\mathcal{a}}, \mathcal{S}]$$

is a surjective homomorphism and therefore

$$[D] \prec [E] \prec [Q^+]$$

is a partial presentation for the initial segment of $P(V_{\mathcal{a}}, \mathcal{S})$. By Theorem 4.10 and the proof of Theorem 4.7, $[E] = [Q^+]$. If S is closed and $(Q^+)^t \subset S$ for some t in $(0,1]$ then we must have $[S] = [R^+]$. This shows that if

$$[Q^+] \prec [S] \prec [R^+]$$

and $[Q^+] \neq [S]$ then $S = R^+$. The proof of the theorem follows from these remarks.

We will consider commutative rings with and without identity. Notation: if R is a ring and $A \subset R$, $B \subset R$ are subsets, let

$$A.B = \{ab\colon a \in A \text{ and } b \in B\} .$$

Theorem 4.12 A topological ring R is $\mathcal{P}_{21}$-metrizable if and only if there exists a denumerable base $(U_n)_{n\in\mathbb{N}}$ for the neighborhoods of 0 satisfying $R.U_n \subset U_n$ and $U_n = -U_n$ for each n in $\mathbb{N}$.

Proof Let $f: R \to \mathbb{R}^+$ satisfy $\mathcal{P}_{21}$ and generate the topology. Let $U_n = \{x \in R : f(x) < 3^{-n}\}$ for n in $\mathbb{N}$. Because f satisfies axiom 5 we have $R.U_n \subset U_n$ for each n. Conversely let $(U_n)_{n\in\mathbb{N}}$ be a base for the neighborhoods of 0 satisfying $U_n = -U_n$ and $R.U_n \subset U_n$. We may also assume that $U_n + U_n + U_n \subseteq U_{n-1}$ for $n \geq 2$. Then if x is in R let

$$f(x) = \text{glb}\ \{\sum_{i=1}^{m} 2^{-n_i} : x \in U_{n_1} + \ldots + U_{n_m}\}.$$

Then f generates the topology of R and satisfies Axioms 1, 2, and 3. We will prove that f also satisfies Axiom 7. Let x in R be given and let y be any other element of R. Given $\varepsilon > 0$ there exist $\{x_1, \ldots, x_m\}$ in R such that

$$x = \sum_{i=1}^{m} x_i\ ,\ x_i \in U_{n_i} \text{ for each } i \text{ and } \sum_{i=1}^{m} 2^{-n_i} < f(x) + \varepsilon\ .$$

Then $xy = \sum_{i=1}^{m} yx_i \in \sum_{i=1}^{m} R.U_{n_i} \subset \sum_{i=1}^{m} U_{n_i}$ and so

$$f(xy) \leq \sum_{i=1}^{m} 2^{-n_i} < f(x) + \varepsilon\ .$$ Therefore $f(xy) \leq f(x)$.

Because x and y are arbitrary elements of R, $f(xy) \leq f(y)$ and finally $f(xy) \leq \min\{f(x), f(y)\}$ and f satisfies Axiom 7.

<u>Theorem 4.13</u> Let $[S] = [S, V_{21}, \mathcal{H}]$ for each S in $\mathcal{H}$. Then

$$[0] < [D] < [W] < [Q^{+}] < [I]$$

is a presentation for $P(V_{21}, \mathcal{H})$ where the first four terms form the initial segment and $[I]$ is the maximum class.

<u>Proof</u> (a) If $[0] < [S]$ and $[0] \neq [S]$ then $[D] < [S]$: because $\{0\} \neq S$ there exists a $\delta > 0$ such that $D \subset \delta.S$. But then, because $[\delta.S] = [S]$, $[D] < [S]$.

(b) If $[D] < [S]$ and $[D] \neq [S]$ then $[W] < [S]$: if 0 is not in $\mathrm{cl}(S - \{0\})$ and the ring R admits a topology generating function $f: R \to S$ satisfying $\mathcal{P}_{21}$ then the topology is discrete and hence

$$g(x) = \begin{cases} 1 & \text{if } x \neq 0 \text{, and} \\ 0 & \text{if } x = 0 \end{cases}$$

generates the topology, satisfies $\mathcal{P}_{21}$ and has values in D. Therefore, because $[D] \neq [S]$, we must have $0 \in \mathrm{cl}(S - \{0\})$ and there exists a strictly decreasing sequence $\{a_n\} \subset S$. We will prove that $W \prec \{a_n\}$.

Let R be a topological ring and let $f: R \to W$ satisfy $\mathcal{P}_{21}$ and generate the topology. Then necessarily f satisfies $\mathcal{P}_{23}$. Define a function $v: R - \{0\} \to \mathbb{N}$ by setting

$$f(x) = 3^{-v(x)} .$$

Then $v(x+y) \geq \min \{v(x), v(y)\}$ and $v(xy) \geq \max \{v(x), v(y)\}$ whenever both sides of the inequalities make sense. Let

$$h(x) = \begin{cases} a_{v(x)} & \text{when } x \neq 0 \text{, and} \\ 0 & \text{when } x = 0 . \end{cases}$$

Then h generates the topology, satisfies $\mathcal{P}_{23}$, and hence $\mathcal{P}_{21}$. This proves that $W \prec \{a_n\}$ and so $[W] \prec [S]$.

(c) If $[W] \prec [S]$ and $[W] \neq [S]$ then $[\mathbb{Q}^+] \prec [S]$: We will consider two cases:

Case I: There exists a $b > 0$ such that $[0,b] \subset cl(S)$. Then by Theorem 1.30 there exists a function g in $\hom(\mathbb{Q}^+, S)$. Let R be a topological ring and let $f: R \to \mathbb{Q}^+$ generate the topology and satisfy the Axioms 1, 2, 3 and 5. Then

$$g \circ f : R \to S$$

satisfies 1, 2 and 3 and generates the topology. We will

see that it also satisfies 5:

If x and y are in R then $f(xy) \leq f(x)$ because f satisfies 5. Then $gf(xy) \leq gf(x)$ because g is increasing. Similarly $gf(xy) \leq gf(y)$ and therefore $gf(xy) \leq \min \{gf(x), gf(y)\}$. We have shown that in Case I $[\mathbb{Q}^+] \prec [S]$.

Case II: The $cl(S)$ is not a neighborhood of 0 in $[0,\infty)$: let R be a topological ring and let $f: R \to S$ be a function generating the topology and satisfying axioms 1, 2, 3 and 5. There exists a monotonically decreasing sequence $\{a_n\}$ in $(0,\infty)$ with limit zero such that the family of sets $W_n = \{x \in R : f(x) < a_n\}$ satisfies $W_{i+1} + W_i = W_i$ for each i in $\mathbb{N}$. (Compare the proof of Theorem 2.4.) For each s in $\mathbb{N}$ let $s.W_n = W_n + \ldots + W_n$ (s terms) and then let $L_n = \bigcup_{s=1}^{m} s.W_{n+1}$. It is easy to check that, for each n,

(i) $R.L_n \subset L_n$,

(ii) $W_{n+1} \subset L_n \subset W_n$,

(iii) $L_n + L_n = L_n$, and

(iv) $L_n = -L_n$.

Let $g: R \to W$ be defined by $g(x) = \text{glb}\,\{3^{-n} : x \in L_n\}$. Then g generates the topology of R and satisfies Axioms

1, 2, 3 and 5 by (i), (ii) and (iii) and the proof of Theorem 4.12. Because g has values in W we have shown that $[S] < [W]$. But by hypothesis $[W] < [S]$ and $[W] \neq [S]$. Thus Case II is impossible and we must have $[Q^+] < [S]$ for all such $[S]$.

(d) For each S in $\mathcal{H}$, $[S] < [I]$: if $f: R \to S$ satisfies $\mathcal{P}_{21}$ and generates the topology then $g(x) = \min \{f(x), 1\}$ generates the topology, satisfies $\mathcal{P}_{21}$, and takes values in I.

(e) The finite field with two elements may be used to show that $[0] \neq [D]$.

(f) The ring $\mathbb{Z}^{\mathbb{N}}$ of Theorem 4.2 (f) may be used to show that $[D] \neq [W]$.

(g) Let $R = \{\begin{pmatrix} 0 & x \\ 0 & 0 \end{pmatrix} : x \in Q\}$ be a subring of the ring of matrices $M(Q,2)$. For each x write $x' = \begin{pmatrix} 0 & x \\ 0 & 0 \end{pmatrix}$ and define $f: R \to Q^+$ by

$$x' \to |x| .$$

Then $x'y' = 0'$ for each x and y in Q and so $f(x'y') \leq f(x')$. Clearly f satisfies Axioms 1, 2, 3 and 5 and generates the usual (subspace of $M(Q,2)$) topology on R. The completion of R is isomorphic to $K = \{x' : x \in R\}$ and therefore is connected. This shows that R is not W-metrizable and therefore $[Q^+] \neq [W]$.

(h) The ring K defined in (g) is I-metrizable but not $\mathbb{Q}^+$-metrizable because it is connected. This shows that $[I] \neq [\mathbb{Q}^+]$ and completes the proof of the theorem.

Theorem 4.14 Let $[S] = [S, V_{21}, \mathcal{H}]$ for each S in $\mathcal{H}$ and let R be a topological ring. Then

(a) R is [D]-metrizable if and only if $\{0\}$ is open in R,

(b) R is [W]-metrizable if and only if there exists a base $(V_n)_{n \in \mathbb{N}}$ for the neighborhoods of 0 satisfying $R.V_n \subset V_n$, $V_n = -V_n$ and $V_n + V_n = V_n$ for all n in $\mathbb{N}$.

Proof (a) is immediate and (b) follows from the proof of Theorem 4.13.

Problem 5 Characterize the property "$[\mathbb{Q}^+, V_{21}, \mathcal{H}]$ - metrizable topological ring" in a similar manner.

Theorem 4.15 A topological ring R is $\mathcal{P}_{22}$-metrizable if and only if there exists a base $(V_n)_{n \in \mathbb{N}}$ for the neighborhoods of 0 satisfying

(i) $V_n \cdot V_m \subset V_{n+m}$ for all n and m in $\mathbb{N}$,

(ii) $V_n + V_n + V_n \subset V_{n-1}$ for all $n \geq 2$ in $\mathbb{N}$, and

(iii) $V_n = -V_n$ for each n in $\mathbb{N}$.

<u>Proof</u> Let $f: R \to R$ generate the topology and satisfy $\mathcal{P}_{22}$. Let

$$V_n = \{x \in R : f(x) < 3^{-n}\} .$$

Then $(V_n)_{n \in \mathbb{N}}$ has the desired properties. Conversely suppose that there exists such a family of subsets of R. For each x in R let

$$f(x) = \mathrm{glb}\{ \sum_{i=1}^{m} 2^{-n_i} : x \in V_{n_1} + \ldots + V_{n_m} \} .$$

Then, by (ii) and (iii), f satisfies axioms 1, 2 and 3 and generates the topology on R. Let x and y in R and $\varepsilon > 0$ be given. Then there exist $V_{n_1}, \ldots, V_{n_m}$ and $V_{l_1}, \ldots, V_{l_k}$ in the given base such that

$x \in V_{n_1} + \ldots + V_{n_m}$ and $y \in V_{l_1} + \ldots + V_{l_k}$ and
$f(x) \leq \sum_{i=1}^{m} 2^{-n_i} < f(x) + \varepsilon$ and $f(y) \leq \sum_{j=1}^{k} 2^{-l_j} < f(y) + \varepsilon$.

Then $xy \in (V_{n_1} + \ldots + V_{n_m}).(V_{l_1} + \ldots + V_{l_k}) \subset \sum_{i,j} V_{n_i}.V_{l_j}$ which, by (i), is contained in $\sum_{i,j} V_{n_i+l_j}$. Thus

$$f(xy) \leq \sum_{i,j} 2^{-n_i-l_j} = (\sum_i 2^{-n_i}).(\sum_j 2^{-l_j}) < (f(x)+\varepsilon)(f(y)+\varepsilon) .$$

This implies that $f(xy) \leq f(x)f(y)$ for all x and y in R and we have proved that f satisfies axiom 5. This completes the proof.

<u>Exercise 4.16</u> Show that $P(V_{22}, \mathcal{H})$ is not linearly ordered, that is, find subsets S and T in $\mathcal{H}$ with $[S] < [T]$ and $[T] < [S]$ both false.

<u>Problem 6</u> Classify all functions on $\mathbb{Q}$ (or $\mathbb{R}$ or $\mathbb{C}$) which satisfy the properties $\mathcal{P}_{22}$ and which generate the usual topology. Note that a very simple complete classification exists when $\mathcal{P}_{22}$ is replaced by $\mathcal{P}_{31}$ or $\mathcal{P}_{32}$, Bourbaki [9, Proposition 4, page 406].

<u>Remarks Related to Problem Six</u>

Let the domain space be $\mathbb{R}$. Let P be the set of functions on $\mathbb{R}$ which satisfy $\mathcal{P}_{22}$ and generate the usual topology. For example the function $x \to |x|$ is in P . It is easy to check that

(a) If f and $g \in P$ and f is increasing on $\mathbb{R}^+$, then $f \circ g \in P$,

(b) If f and $g \in P$ then $f + g \in P$,

(c) If $f \in P$ and $\beta \geq 1$ then $\beta.f \in P$ where $(\beta.f)(x) = \beta.f(x)$,

(d) If $f \in P$ and $0 < \alpha \leq 1$ then $f^\alpha \in P$ where $f^\alpha(x) = (f(x))^\alpha$.

Using these properties we can deduce that, for example,

$$f(x) = (4(|x|^{19/23} + 3|x|^{1/2})^{3/7} + 11(14|x|^{5/11} + 2|x|^{3/4})^{2/3})$$

is in P . Let $D \subset P$ be the set of functions which can be obtained by applying the operations (a), (b), and (c) finitely many times to the function $x \to |x|$. One might think that $D = P$. To see that this is false let

$$h(x) = \begin{cases} 2x & \text{when } 0 \leq x \leq \frac{1}{2}, \\ 1 & \text{when } \frac{1}{2} \leq x \leq 1, \\ x & \text{when } 1 \leq x . \end{cases}$$

Then $h \in P$ and the derivative $h'(3/4) = 0$. If $f \in D$ then $f'(x) > 0$ for each $x > 0$ in R . Therefore h is not in D and so D is not equal to P .

We list here several properties common to all functions f in P . Some are consequences of Axiom 3 for rings (subadditivity), some of Axiom 6 (submultiplicativity) and some consequences of both these axioms.

<u>Property 4.17</u> Let $\beta > 0$ and $\lambda > 1$ be given. For each $\alpha > 0$ the inequality $f(n\alpha) \geq \beta(n\alpha)^{\lambda}$ is true for at most a finite number of integers n in $\mathbb{N}$.

Proof Suppose there exists a subsequence $(n_i)_{i \in \mathbb{N}}$ of $\mathbb{N}$ such that

$$f(n_i \alpha) \geq \beta (n_i \alpha)^{\lambda}$$

for all i in $\mathbb{N}$. Then, by Axiom 3, for all i

$$n_i f(\alpha) \geq f(n_i \alpha) \geq \beta n_i^{\lambda} \alpha^{\lambda}$$

and thus

$$f(\alpha) \geq \beta \alpha^{\lambda} . n_i^{\lambda - 1} .$$

This is impossible because $f(\alpha) < \infty$.

Property 4.18 For all $\alpha > 0$ and $\lambda > 1$

$$\lim_{n \to \infty} \frac{f(n\alpha)}{(n\alpha)^{\lambda}} = 0 .$$

Proof Given $\varepsilon > 0$, $f(n\alpha) \geq \varepsilon (n\alpha)^{\lambda}$ is false for all but finitely many n in $\mathbb{N}$ by Property 4.17. Thus there exists an $N = N(\varepsilon)$ in $\mathbb{N}$ such that $f(n\alpha) < \varepsilon (n\alpha)^{\lambda}$ for all $n \geq N$, that is $f(n\alpha)/(n\alpha)^{\lambda} < \varepsilon$. This proves the property.

Property 4.19 For all $\lambda > 1$ $\lim_{x \to \infty} f(x).x^{-\lambda} = 0$.

Proof If f is in P then f is uniformly continuous on $[0,\infty)$ (compare Theorem 1.34). Given $\varepsilon > 0$ there is a $\delta = \delta(\varepsilon) > 0$ such that if $|x - y| < \delta$ then

$|f(x) - f(y)| < \varepsilon$. Choose α so that $0 < \alpha < \delta$; $\alpha = \alpha(\varepsilon)$. Let $0 < x < \infty$ be given. Chose $n = n(\alpha, \varepsilon, x) = n(\varepsilon, x)$ in $\mathbb{N}$ such that $x - \delta < n\alpha \leq x$.

Given ε and α there exists an $N = N(\varepsilon, \alpha) = N(\varepsilon)$ in $\mathbb{N}$ such that $f(n\alpha) \leq \varepsilon(n\alpha)^{\lambda}$ when $n \geq N$ by Property 4.18. Let
$M = M(\varepsilon) = \max \{N(\varepsilon).\alpha(\varepsilon) + \delta(\varepsilon), 1 + \delta(\varepsilon)\}$. Then if we chose $x \geq M$ we have $x \geq N.\alpha + \delta$ and so $n.\alpha > x - \delta \geq N.\alpha$, that is to say $n > N$. This shows that $f(n\alpha).(n\alpha)^{-\lambda} < \varepsilon$. We also have
$|f(x) - f(n\alpha)| < \varepsilon$. Because $x \geq M$ we have $x \geq 1 + \delta$ or $x - \delta \geq 1$, hence $(n\alpha)^{-\lambda} \leq 1$. Then

$$f(x)x^{-\lambda} \leq |f(x)x^{-\lambda} - f(n\alpha)x^{-\lambda}| + |f(n\alpha)x^{-\lambda} - f(n\alpha)(n\alpha)^{-\lambda}| + f(n\alpha)(n\alpha)^{-\lambda}$$ and therefore

$f(x)x^{-\lambda} \leq |f(x) - f(n\alpha)|\ x^{-\lambda} + 3f(n\alpha)(n\alpha)^{-\lambda}$ because $x \geq n\alpha$, and so $\leq 1/4\varepsilon + 3/4\varepsilon = \varepsilon$.

The proof is complete because this inequality holds for all $x \geq M = M(\varepsilon)$.

<u>Property 4.20</u> If $x \neq 0$ then
$1 \leq f(1) \leq f(x).f(x^{-1})$ by Axiom 6.

Property 4.21 If f is in P then f is unbounded.

Proof Let $M \geq 1$ be given. Because $f(0) = 0$, $f(1) \geq 1$ and f is continuous there exists an x in $(0,1]$ with $f(x) = M^{-1}$. By Property 4.20 $M = f(x)^{-1} \leq f(x^{-1})$. This proves that f is not bounded on R.

Property 4.22 If f is in P and there exist n numbers $(a_i)_{1 \leq i \leq n}$ in R such that $f(x) = a_1 x + \ldots + a_n x^n$ with $a_n \neq 0$ then $n = 1$.

Proof Because f is positive and unbounded $a_n > 0$ by Property 4.21. If $n \neq 1$ let $\lambda = n - \frac{1}{2}$. Then $\lim_{x \to \infty} f(x)x^{-\lambda} = \infty$, a contradiction by Property 4.19. Thus we must have $n = 1$.

Property 4.23 If q is a polynomial with coefficients in $\mathbb{N}$ then $f(q(x)) \leq q(f(x))$ for all x in R and f in P.

Property 4.24 Let $\lambda > 0$ and let f in P be such that $f(x) \geq x^{\lambda}$ for $0 \leq x \leq \delta$ where $\delta > 0$. Then $f(x) \geq x^{\lambda}$ on $[0,1]$.

Proof Assume that $0 < \delta < 1$ and $0 < x < 1$. Note that $f(1) \geq 1 = 1^{\lambda}$ necessarily. There exists an n in $\mathbb{N}$ such that $0 < x^n < \delta$ and therefore

$f(x)^n \geq f(x^n) \geq (x^n)^\lambda = (x^\lambda)^n$. Thus $f(x) \geq x^\lambda$.

<u>Property 4.25</u> If there exists a sequence $x_n > 0$ with limit zero such that $f(x_n) \leq x_n$ for all n then $f(x) = |x|$ on $\mathbb{R}$.

<u>Proof</u> Let $\varepsilon > 0$ and $x > 0$ be given. There exists an $x_n < \varepsilon$ and an m in $\mathbb{N}$ such that $x \leq mx_n < x + \varepsilon$. Then, when f is increasing,

$$f(x) \leq f(mx_n) \leq mf(x_n) \leq mx_n < x + \varepsilon$$

and so $f(x) \leq x$. If not we give an alternative argument: there exists $\delta_1 > 0$ such that $|x - y| < \delta_1$ implies $|f(x) - f(y)| < \frac{1}{2}\varepsilon$. Let $\delta_2 = \min\{\frac{1}{2}\varepsilon, \delta_1\}$. Choose n in $\mathbb{N}$ so that $0 < x_n < \delta_2$ and m in $\mathbb{N}$ so that $|mx_n - x| < \delta_2$. Then $f(x) < f(mx_n) + \frac{1}{2}\varepsilon \leq mx_n + \frac{1}{2}\varepsilon < x + \varepsilon$. Because ε is arbitrary we have $f(x) \leq x$ for all $x > 0$. If there exists a y with $0 < f(y) < y$ then $y^{-1} < f(y)^{-1} \leq f(y^{-1})$ by Property 4.20. This is impossible and hence $f(x) = x$ for all $x > 0$. Thus $f(x) = x$ for all $x \geq 0$ from which the conclusion follows.

<u>Property 4.26</u> If for some $\delta > 0$ $f(x) = x$ on $[0,\delta]$, then $f(x) = x$ on $[0,\infty)$.

<u>Proof</u> An immediate consequence of Property 4.25.

<u>Property 4.27</u> If $\lambda > 0$ is such that $f(x) \geq \lambda x$ for all x then $f(x) \geq x$ for all x.

<u>Proof</u> By Axiom 6 $f(x)^n \geq \lambda x^n$ and so $f(x) \geq \lambda^{1/n} x$ for all n. Therefore $f(x) \geq x$.

We return now to the study of metrizable topological rings and fields.

<u>Theorem 4.28</u> A topological ring R is $\mathcal{P}_{23}$-metrizable if and only if there exists a denumerable base $(U_n)_{n\in \mathbb{N}}$ for the neighborhoods of 0 satisfying

(i) $R.U_n \subset U_n$,

(ii) $U_n + U_n = U_n$, and

(iii) $U_n = -U_n$ for all n.

<u>Proof</u> Let $f: R \to R^+$ satisfy $\mathcal{P}_{23}$ and generate the topology of R. For each n in $\mathbb{N}$ let $U_n = \{x \in R : f(x) < 3^{-n}\}$. The family $(U_n)_{n\in \mathbb{N}}$ is a base for the neighborhoods of zero and satisfies (i), (ii) and (iii). Conversely let such a base $(U_n)_{n\in \mathbb{N}}$ be given.

Define

$$f(x) = \min \{3^{-n} : x \in U_n\}$$

for each x in R . The function f has the desired properties.

<u>Theorem 4.29</u> A topological ring R is $\mathcal{P}_{24}$-metrizable if and only if there exists a denumerable base $(U_n)_{n\in\mathbb{N}}$ for the neighborhoods of 0 satisfying

(i) $U_n . U_m \subset U_{n+m}$ for all n and m in $\mathbb{N}$,

(ii) $U_n + U_n = U_n$ for each n in $\mathbb{N}$, and

(iii) $U_n = -U_n$ for each n .

<u>Proof</u> This is left as an exercise; compare with Theorem 4.15.

<u>Lemma 4.30</u> Let $(a_n)_{n\in\mathbb{N}}$ be a strictly monotonically decreasing sequence with $a_1 < 1$ and limit zero. Then there exists a subsequence $(a_{n_j})_{j\in\mathbb{N}}$ such that $a_{n_{i+j}} \leq a_{n_i} . a_{n_j}$ for each i and j in $\mathbb{N}$.

<u>Proof</u> Let $a_{n_1} = a_1$. Suppose $a_{n_1}, \ldots, a_{n_k}$ have been defined so that $a_{n_{i+j}} \leq a_{n_i} . a_{n_j}$ whenever $i + j \leq k$. Note that the condition holds vacuously when $k = 1$. Choose n_{k+1} so that

$a_{n_{k+1}} \leq \min \{a_{n_l} . a_{n_m} : l \in \mathbb{N} , m \in \mathbb{N}$ and $l + m = k + 1\}$.
Then $n_{k+1} < n_k$ and the subsequence $(a_{n_i})_{i \in \mathbb{N}}$ has the required properties.

Let $\frac{1}{2} D = \{0,\frac{1}{2}\} \in \mathcal{H}$.

Theorem 4.31 For each S in $\mathcal{H}$ let $[S] = [S,V_{24},\mathcal{H}]$. Then

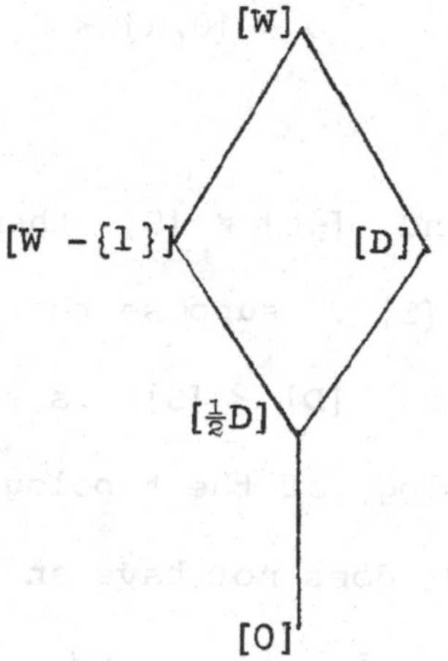

is the graph of a complete presentation for $P(V_{24},\mathcal{H})$.

Proof We will prove the theorem step by step.

(a) Let $S \in \mathcal{H}$. Then $[0] \prec [S] \prec [W]$: this follows from the proof of Theorem 4.29.

(b) If $[0] \prec [S]$ and $[0] \neq [S]$ then $[\frac{1}{2} D] \prec [S]$: since $S \neq \{0\}$ there is an $a > 0$ with $\{0,a\} \subset S$. If

R is a topological ring and $f: R \to \frac{1}{2}D$ a function satisfying ρ_{24} and generating the topology then, if x and y are in R, $f(xy) \leq \frac{1}{2}.\frac{1}{2} < \frac{1}{2}$ and therefore $xy = 0$. Let $g: R \to \{0,a\}$ be defined by

$$g(x) = \begin{cases} a & \text{if } x \neq 0 \text{, and} \\ 0 & \text{if } x = 0 \text{.} \end{cases}$$

Then g satisfies axioms 1, 2 and 4, generates the topology, and satisfies axiom 6 because $xy = 0$ for all x and y in R. Therefore $\frac{1}{2}D \prec \{0,a\} \prec S$ and so $[\frac{1}{2}D] \prec [S]$.

(c) If $[\frac{1}{2}D] \prec [S]$ and $[\frac{1}{2}D] \neq [S]$ then either $[D] \prec [S]$ or $[W - \{1\}] \prec [S]$: suppose that $[\frac{1}{2}D] \prec [S]$, $[\frac{1}{2}D] \neq [S]$ are both true and $[D] \prec [S]$ is false. Let $f: R \to S$ generate the topology of the topological ring R and satisfy ρ_{24}. Then R does not have an identity. (If so $f(1) \leq f(1)^2$ and so $1 \leq f(1)$ and $\{0, f(1)\} \subset S$. From $D \prec \{0, f(1)\}$ we obtain a contradiction.) Because $[\frac{1}{2}D] \neq [S]$ there exists a strictly monotonically decreasing sequence (a_n) with $0 < a_n < 1$ for all n and limit zero. There exists a subsequence $(a_{n_j})_{j \in \mathbb{N}}$ with $a_{n_{i+j}} \leq a_{n_i}.a_{n_j}$ for all i and j by Lemma 4.30. Let K be a topological ring and $f: K \to W - \{1\}$ a function satisfying ρ_{24} and generating the topology of K. If

$x \in K$ and $x \neq 0$ let

$$g(x) = 3^{-v(x)} .$$

Then $v(x+y) \geq \min \{v(x), v(y)\}$ and $v(xy) \geq v(x) + v(y)$ whenever both sides of the inequalities make sense. Let

$$h(x) = \begin{cases} a_{n_{v(x)}} & \text{when } x \neq 0 \text{, and} \\ 0 & \text{when } x = 0 . \end{cases}$$

The function $h: K \to S$ generates the topology and satisfies $\mathcal{P}_{24}$. This shows that $[W - \{1\}] < [S]$.

(d) $[\frac{1}{2}D] \neq [0]$: the ring of continuous real valued functions on R with compact support and the discrete topology is $\frac{1}{2}D$ - metrizable but not $\{0\}$ - metrizable.

(e) $[W - \{1\}] \neq [\frac{1}{2}D]$: let $\mathbb{Q}_3$ be the 3-adic completion of the rational numbers and let $R = \{x \in \mathbb{Q}_3 : |x|_3 < 1\}$. Then R with the induced topology is a topological ring, $|\cdot|_3$ restricted to R satisfies $\mathcal{P}_{24}$, generates the topology, and has values in $W - \{1\}$. R is not $\frac{1}{2}D$ - metrizable because its topology is not discrete.

(f) $[D] \neq [\frac{1}{2}D]$: the finite field $\mathbb{Z}_2$ with the discrete topology is $[D, V_{24}, \mathcal{H}]$-metrizable and not $[\frac{1}{2}D, V_{24}, \mathcal{H}]$-metrizable.

(g) $[W - \{1\}] \neq [W]$: let $R = \{x \in Q_3 : |x|_3 \leq 1\}$. Then R is W-metrizable and not $\{W - \{1\}\}$-metrizable.

(h) $[D] \neq [W]$: consider the example given in (g) and note that the induced topology is not discrete.

(i) $[D] \neq [W - \{1\}]$: consider the example given in (e).

To complete the proof we need only observe that $0 < \frac{1}{2}D$, $\frac{1}{2}D < W - \{1\}$, $\frac{1}{2}D < W$, $W - \{1\} < W$ and $D < W$.

Theorem 4.32 Let [S] denote the class $[S, V_{23}, \mathcal{H}]$ for each S in $\mathcal{H}$. Then

$$[0] < [D] < [W]$$

is a complete presentation for $P(V_{23}, \mathcal{H})$.

Proof (a) If S is in $\mathcal{H}$ then $[0] < [S] < [W]$ by the proof of Theorem 4.28.

(b) If $[0] < [S]$ and $[0] \neq [S]$ then $[D] < [S]$: because $0 \neq S$ there exists an $a > 0$ in S . If $f: R \to D$ is a function on a topological ring satisfying ρ_{23} and generating the topology then $a.f: R \to S$ has these same properties. Thus $[D] < [S]$.

(c) $[0] \neq [D]$: see example (f) of Theorem 4.31.

(d) [D] $\neq$ [W] : see example (g) of Theorem 4.31 and note that $|xy|_3 \leq \min\{|x|_3, |y|_3\}$ on R . The proof is complete.

It is natural to study in this manner topological rings with real valued functions satisfying, say, properties $\mathcal{P}_{31}$. Any such ring R is necessarily an integral domain and we may form the field of quotients F_R . As is well known a function f on R satisfying $\mathcal{P}_{31}$ extends to a function $\hat{f}$ on F_R which satisfies these same axioms.

<u>Theorem 4.33</u> Let $\mathcal{S} = \{S \in \mathcal{H} : t.S \subset S \text{ for all } t \in S\}$. Let $\mathcal{A}$ be the class of $\mathcal{P}_{31}$-metrizable topological fields and let $\mathcal{B}$ be the class of $\mathcal{P}_{31}$-metrizable topological rings with datums $V_{\mathcal{A}}$ and $V_{\mathcal{B}}$ respectively. Then $P(V_{\mathcal{A}}, \mathcal{S}) = P(V_{\mathcal{B}}, \mathcal{S})$.

<u>Proof</u> Let $[S, V_{\mathcal{A}}] = [S, V_{\mathcal{A}}, \mathcal{S}]$ and $[S, V_{\mathcal{B}}] = [S, V_{\mathcal{B}}, \mathcal{S}]$ for all $S \in \mathcal{S}$. If $[S, V_{\mathcal{B}}] < [T, V_{\mathcal{B}}]$ then $[S, V_{\mathcal{A}}] < [T, V_{\mathcal{A}}]$ because $\mathcal{A} \subset \mathcal{B}$. This shows that the mapping

$$[S, V_{\mathcal{B}}] \to [S, V_{\mathcal{A}}]$$

$$g: P(V_{\mathcal{B}}, \mathcal{S}) \to P(V_{\mathcal{A}}, \mathcal{S})$$

is a well defined homomorphism. It is easy to see that g

is onto. Now let $[S,V_{\mathcal{A}}] < [T,V_{\mathcal{A}}]$ and suppose that R is a topological ring with $f\colon R \to S$ a function satisfying $\mathcal{P}_{31}$ and generating the topology. Then if $[r/s]$ is in F_R, the extension $\hat{f}([r/s]) = f(r)/f(s)$ is in S and $\hat{f}\colon F_R \to S$ satisfies $\mathcal{P}_{31}$. Because $[S,V_{\mathcal{A}}] < [T,V_{\mathcal{A}}]$ there exists a function $k\colon F_R \to T$ satisfying $\mathcal{P}_{31}$ and generating the same topology as $\hat{f}$. Then the map k when restricted to $R \subset F_R$ generates the same topology on R as $\hat{f}$ restricted to R, that is, the given topology on R, takes values in T and satisfies $\mathcal{P}_{31}$. This proves that $[S,V_{\mathcal{B}}] < [T,V_{\mathcal{B}}]$ and completes the proof that g is an isomorphism.

<u>Remark 4.34</u> The morphisms for the categories topological rings and fields are, as usual, continuous homomorphisms. It is easy to see that for each of these categories and for each related $\mathcal{P}_{ij}$ considered above the property $\mathcal{P}_{ij}$ is natural for the given category, the property "$[S,V_{ij},\mathcal{S}]$-metrizable" is intrinsic by Theorem 1.6, and the classes $[S,V_{ij},\mathcal{S}]$ are invariants.

Topological Vector Spaces

We are interested in metrizable topological vector spaces over specific topological fields. The fields are $\mathbb{C}$, $\mathbb{R}$, and $\mathbb{Q}$ with the usual absolute value or modulus (satisfying $\mathcal{P}_{31}$) , $\mathbb{Q}_p$ for primes p with functions equivalent to the p-adic valuation (satisfying $\mathcal{P}_{32}$) , and $\mathbb{F}_q$, $q = p^n$, the finite fields with discrete topology and topology generating function $f(x) = 1$ for all non zero x in $\mathbb{F}_q$. The symbol $\mathbb{K}$ will denote one of these fields and the symbol $\mathcal{V}$ the family of all metrizable topological vector spaces over these fields.

Metric like functions on the topological vector spaces will have real values and satisfy a set of axioms. If $\mathbb{K}$ is one of the given fields we will represent its topology generating functions by $|\cdot|$. Note that $|\cdot|$ satisfies the properties $\mathcal{P}_{31}$ in every case.

<u>Theorem</u> (Horváth [30, page 111]) Let $(V,\mathcal{T})$ be a topological vector space over a field $\mathbb{K}$. Then there exists a function $f: V \to \mathbb{R}$ satisfying $\mathcal{T} = \mathcal{T}_f$ and

(1) $f(x) = 0$ if and only if $x = 0$,

(2) $f(-x) = f(x)$ for all x ,

(3) $f(x+y) \leq f(x) + f(y)$ for all x and y ,

(8) if $a \to 0$ in $\mathbb{K}$ and $x \in V$ then $f(ax) \to 0$ in $\mathbb{R}$.

(9) if $a \in \mathbb{K}$ and $|a| \leq 1$ then $f(ax) \leq f(x)$ for all x .

Definition 4.35 Let $\mathcal{P}_{41}$ be the conjunction of the above properties, let $\mathcal{R}$ be the usual rule for generation of TVS topologies and let $V_{41} = (\mathcal{V}, \mathcal{P}_{41}, \mathcal{R})$.

Theorem 4.36 For each S in $\mathcal{H}$ let $[S] = [S, V_{41}, \mathcal{H}]$. Then

$$[0] < [D] < [W] < [\mathbb{Q}^+] < [I]$$

is a presentation for $P(V_{41}, \mathcal{H})$ in which the first four terms form the initial segment and $[I]$ is the maximum class.

Proof (a) If S in $\mathcal{H}$ is such that $[S] \neq [0]$ then $[D] < [S]$ because if f satisfies $\mathcal{P}_{41}$ so does $a.f$ for all $a > 0$ in $\mathbb{R}$.

(b) If $S \in \mathcal{H}$ is such that $cl(S - \{0\})$ does not contain 0 then $[S] < [D]$.

(c) If S in $\mathcal{H}$ is such that $0 \in cl(S - \{0\})$ then $[W] < [S]$: there exists a sequence $\{a_n\} \subset S$ of distinct points with $a_n \to 0$. If $(V, \mathcal{T})$ is a TVS and $f: V \to W$ satisfies $\mathcal{P}_{41}$ and generates the topology consider V as a topological group. By Theorem 4.2 there exists a function $g: V \to \{a_n\}$ satisfying axioms 1, 2 and 3 and $\mathcal{T} = \mathcal{T}_f = \mathcal{T}_g$. If $a \to 0$ in $\mathbb{K}$ then $ax \to 0$ in $\mathcal{T}$ and therefore $g(ax) \to 0$. Thus g satisfies 8. From the definition of g (Theorem 4.2) if $f(x) \leq f(y)$ then $g(x) \leq g(y)$. Thus $g(ax) \leq g(x)$ if $|a| \leq 1$ because f satisfies 9. Therefore g also satisfies axiom 9.

(d) If $[D] \prec [S]$ and $[D] \neq [S]$ then $[W] \prec [S]$: this follows from (b) and (c).

(e) If $S \in \mathcal{H}$ is such that $cl(S)$ is not a neighborhood of zero in $[0,\infty)$ then $[S] \prec [W]$: let $(V,\mathcal{T})$ be a TVS and $f: V \to S$ a function satisfying $\mathcal{P}_{41}$ and $\mathcal{T}_f = \mathcal{T}$. Regard V as a topological group. Then there exists a function $g: V \to W$ satisfying Axioms 1, 2 and 3 and $\mathcal{T} = \mathcal{T}_g$. By the same argument as in (c) g satisfies axiom 8. For each x in V let

$$h(x) = \max \{g(ax) : |a| \leq 1\} .$$

Then $h(ax) \leq h(x)$ for all a in $\mathbb{K}$ with $|a| \leq 1$, $h: V \to W$, and h satisfies axioms 1, 2, 3 and 8. We will prove that $\mathcal{T} = \mathcal{T}_h$; given $\varepsilon > 0$ there exists a $\delta > 0$ such that if $f(x) < \delta$ then $g(x) < \varepsilon$. Let a in $\mathbb{K}$ satisfy $|a| \leq 1$. Because f satisfies axiom 9, $f(ax) \leq f(x) < \delta$ and so $g(ax) < \varepsilon$. Therefore $h(x) \leq \varepsilon$ if $f(x) < \delta$. This shows the topology generated by h is coarser than that generated by f and so, because $h(x) \geq g(x)$ for all x, $\mathcal{T} = \mathcal{T}_h$.

(f) If S in in $\mathcal{H}$ is such that $cl(S)$ is a neighborhood of zero in $[0,\infty)$ then $[\mathbb{Q}^+] < [S]$: if f satisfies $\mathcal{P}_{41}$, has range in $\mathbb{Q}^+$ and $g \in hom(\mathbb{Q}^+, S)$ then $g \circ f$ satisfies $\mathcal{P}_{41}$.

(g) If S is in $\mathcal{H}$ then $[S] < [I]$: if f satisfies $\mathcal{P}_{41}$ so does $g = \min\{f,1\}$.

(h) To see that the classes given in the theorem statement are distinct consider the fields $\mathbb{F}_2$, $\mathbb{Q}_3$, $\mathbb{Q}$ and $\mathbb{R}$ as vector spaces over themselves.

The proof follows from these assertions.

<u>Remark 4.37</u> As one would suspect we obtain a different result if we restrict our attention to vector spaces over a fixed field $\mathbb{K}$ in the given set of fields. Let $\mathcal{V}_{\mathbb{K}}$ be the category of $\mathcal{P}_{41}$-metrizable topological vect spaces over the field $\mathbb{K}$ and let

$$V_{\mathbb{K}} = (\mathcal{V}_{\mathbb{K}}, \mathcal{P}_{41}, \mathcal{R}) .$$

Let E be a TVS over $\mathbb{K}$ and suppose that $E \neq \{0\}$. Then the linear map

$$\mathbb{K} \to E$$

$$a \to a.x ,$$

where x is a fixed non zero element of E, is an embedding of $\mathbb{K}$ in E. (The TVS is Hausdorff because it is assumed to be metrizable.) If S in $\mathcal{H}$ is such that E is $[S,V_{\mathbb{K}},\mathcal{H}]$-metrizable then so is $\mathbb{K}$.

If $\mathbb{K}$ is a connected field then necessarily $[S,V_{\mathbb{K}},\mathcal{H}] > [I,V_{\mathbb{K}},\mathcal{H}]$. This proves that

$$[0,V_{\mathbb{K}},\mathcal{H}] < [I,V_{\mathbb{K}},\mathcal{H}]$$

is a complete presentation for $P(V_{\mathbb{K}},\mathcal{H})$ when $\mathbb{K}$ is $\mathbb{C}$ or $\mathbb{R}$. If $\mathbb{K} = \mathbb{Q}$ then $\mathbb{K}$ is not $[S,V_{\mathbb{K}},\mathcal{H}]$-metrizable for any S in $\mathcal{H}$ satisfying $cl(S)$ is not a neighborhood of zero in $[0,\infty)$, and therefore any $[S,V_{\mathbb{K}},\mathcal{H}]$-metrizable TVS must be $\{0\}$. This shows that $[0,V_{\mathbb{Q}},\mathcal{H}] = [W,V_{\mathbb{Q}},\mathcal{H}]$.

Finally we remark that the map $[S,V,\mathcal{H}] \to [S,V_{\mathbb{K}},\mathcal{H}]$ is a surjective, not necessarily injective, homomorphism for each $\mathbb{K}$, where $V = V_{41}$ is defined above.

<u>Remark 4.38</u> If E is a $\mathbb{K}$ vector space let $\dim_{\mathbb{K}} E$ be the algebraic dimension of E over the field $\mathbb{K}$. Let $\mathcal{A}$ be the class of $\mathcal{P}_{41}$-metrizable $\mathbb{K}$ TVS's E which satisfy $\dim_{\mathbb{K}} E \leq \aleph_0$ and let $V'_{\mathbb{K}} = (\mathcal{A},\mathcal{P}_{41},\mathcal{R})$. If $\mathrm{card}(\mathbb{K}) \leq \aleph_0$ and S is in $\mathcal{H}$ then we readily see that $[S,V'_{\mathbb{K}},\mathcal{H}] < [\mathbb{Q}^+,V'_{\mathbb{K}},\mathcal{H}]$. This implies that

$$[0,V'_{\mathbb{Q}},\mathcal{H}] < [\mathbb{Q}^+,V'_{\mathbb{Q}},\mathcal{H}]$$

is a complete presentation for $P(V'_{\mathbb{Q}},\mathcal{H})$.

Chapter V : Generation by dense subspaces

Metrizable topological and uniform spaces

Theorem 5.1 Let $(X,\mathcal{T})$ be second countable and metrizable and let $D \subset X$ be a countable and dense subset. Then there exists a metric d on X with $\mathcal{T}_d = \mathcal{T}$ and $d(D^2) \subset \mathbb{Q}^+$.

Proof Let h be a metric on X compatible with the topology $\mathcal{T}$. Then $\mathrm{card}(h(D^2)) \leq \aleph_0$ and so, by the proof of Theorem 1.30, there exists an f in $\hom(R^+, R^+)$ such that $f(h(D^2)) \subset \mathbb{Q}^+$. Then $d = f \circ h$ is a metric on X, equivalent to h by Theorem 1.8, satisfying $d(D^2) \subset \mathbb{Q}^+$.

The same proof shows

Theorem 5.2 Let $(X,\mathcal{U})$ be a metrizable uniform space with second countable induced topology $\mathcal{T}_{\mathcal{U}}$ and countable dense subset D. Then there exists a metric d on X, generating $\mathcal{U}$ and satisfying $d(D^2) \subset \mathbb{Q}^+$.

It is easy to see that the following stronger result is true: Let (X,d) be a metric space, let $D \subset X$ be countable, and let $S \subset [0,\infty)$ be dense in a neighborhood of zero. Then there exists a metric h on X which is uniformly equivalent to d on X, and which satisfies $h(D^2) \subset S$.

It is not known whether or not Theorems 5.1 and 5.2 can be extended to spaces of arbitrary weight. The following weak result is easily proven:

<u>Theorem 5.3</u> Let $(X,\mathcal{T})$ be metrizable. Then there exists a dense $\mathbb{Q}^+$- metrizable subset $D \subset X$.

<u>Proof</u> To construct the subset D we will use an idea which is a very old one, Comfort [19, page 170]. Let d be a metric on X satisfying $\mathcal{T} = \mathcal{T}_d$. Let D_1 be a maximal subset of X having the property $d(x,y) \geq 1$ when x and y are in D_1 and $x \neq y$. By induction let D_{n+1} be a maximal superset of D_n having the property $d(x,y) \geq 1/(n+1)$ when x and y are in D_{n+1} and $x \neq y$. Then because each D_n is a discrete subset of X it is closed and has $\operatorname{Ind} D_n = 0$. Clearly $(D_n)_{n\in\mathbb{N}}$ is locally countable. Therefore, by the theorem of Kimura [32], D has large inductive dimension zero, where $D = \bigcup_{n=1}^{\infty} D_n$. Thus, by Theorem 3.5, D is H-metrizable and in particular $\mathbb{Q}^+$-metrizable. By [19, page 169], D is dense.

<u>Conjecture 5.4</u> There exists a Q^+-metric on D which extends to a compatible metric on X .

<u>Other Categories</u>

We will now extend these ideas to more general categories. All objects in the categories considered will be sets with additional structure which includes a natural associated topology. Subobjects will be subsets whose algebraic operations, if any, are the restrictions of the algebraic operations in the original space. The natural topology on subobjects is to be the usual subspace topology. Such categories will be called normal. We are interested in the categories studied earlier in this work.

<u>Definition 5.5</u> Let $\mathcal{A}$ be a normal real-generated category, $\mathcal{S}$ a subfamily of $\mathcal{J}$, $\mathcal{P}$ be a property of families of functions on powers of objects in $\mathcal{A}$, which is natural for $\mathcal{A}$, $\mathcal{R}$ the rule for the generation of (at least part of) the structure of objects in $\mathcal{A}$. Let $V = (\mathcal{A}, \mathcal{P}, \mathcal{R})$ be the datum of $\mathcal{A}$.

Let $(X, \mathfrak{T}) \in \mathcal{A}$ and S be in $\mathcal{S}$. We say X is $[S, V, \mathcal{S}]$-generated if there exists a subobject $Y \subset X$ such that $cl(Y) = X$ and a family of functions $F: X^n \to R$

with $\mathfrak{C} = \mathfrak{C}_F$ and

$$[F(Y^n)] < [S]$$

where $F(Y^n) = \cup \{f(Y^n): f \in F\}$. If Y satisfies this condition we say Y is an $[S,V,\mathcal{S}]$-generating subobject of X . For example X is an $[R,V,\mathcal{S}]$-generating subobject of X , if $R \in \mathcal{S}$,and if X is $[S,V,\mathcal{S}]$-metrizable then X is an $[S,V,\mathcal{S}]$-generating subobject of X for all normal real-generated categories.

If $V = (\mathcal{J}, \mathcal{P}_2, \mathcal{R})$ and $(X,\mathcal{T})$ is a separable metrizable space then $(X,\mathcal{T})$ is $[\mathbb{Q}^+,V,\mathcal{H}]$-generated by Theorem 5.1. If $(X,\mathfrak{U})$ is a separable metrizable uniform space then the same is true. Indeed we can say more. Let $\mathcal{E}$ be the class of uniform spaces with structure generated by families of pseudometrics, $V = (\mathcal{E}, \mathcal{P}_{20}, \mathcal{R})$, as described in Chapter II, then

<u>Theorem 5.6</u> Any separable uniform space $(X,\mathfrak{U})$ is $[\mathbb{Q}^+,V,\mathcal{H}]$-generated by any countable dense subset $Y \subset X$. If for some $S \in \mathcal{H}$ every separable uniform space is $[S,V,\mathcal{H}]$-generated then $[\mathbb{Q}^+] < [S]$.

<u>Proof</u> Let $Y \subset X$ be countable and dense and let $F = \{d_j: j \in J\}$ be a family of pseudometrics generating $\mathfrak{U}$. Let, for each j in J , $g_j \in \hom(R^+,R^+)$ be such that $g_j(d_j(Y^2)) \subset \mathbb{Q}^+$ and let $h_j = g_j \circ d_j$ (compare

the proof of Theorem 5.1 above). Then the family $G = \{h_j : j \in J\}$ generates $\mathcal{U}$ and $G: Y^2 \to \mathbb{Q}^+$.

If S in $\mathcal{H}$ is such that every separable uniform space is $[S,V,\mathcal{H}]$-generated then $\mathbb{Q}$ with its usual uniform structure $\mathcal{U}$ has a dense subset $Y \subset \mathbb{Q}$ and a family of pseudometrics $F = \{d_j : j \in J\}$ satisfying $\mathcal{U} = \mathcal{U}_F$ and $[d_j(Y^2)] < [S]$ for each j in J . If $[\mathbb{Q}^+] < [S]$ were false, then by Theorem 2.27, necessarily $[S] < [D]$ and Y as a uniform subspace of X is $[D,V,\mathcal{H}]$-metrizable. This implies that the completion of Y is $[D,V,\mathcal{H}]$-metrizable which is impossible because the completion is uniformly isomorphic to R . Therefore we must have $[\mathbb{Q}^+] < [S]$.

Let V_6 be defined as above in Chapter IV.

<u>Theorem 5.7</u> Let $(G,\mathcal{T})$ be a separable $[I,V_6,\mathcal{H}]$-metrizable topological group. Then $(G,\mathcal{T})$ is $[\mathbb{Q}^+,V_6,\mathcal{H}]$-generated by any countable dense subgroup. If S in $\mathcal{H}$ is such that every such group is $[S,V_6,\mathcal{H}]$-generated then $[\mathbb{Q}^+,V_6,\mathcal{H}] < [S,V_6,\mathcal{H}]$.

<u>Proof</u> This is similar to that of Theorem 5.6 and depends on Theorem 4.2. We need only observe that the

group generated by a countable dense subset is a countable dense subgroup, that is there exist countable dense subgroups.

Theorem 5.8 Let $ij \in \{21, 23, 24\}$ and let $(R, \mathcal{T})$ be a separable $[R^+, V_{ij}, \mathcal{H}]$-metrizable topological ring. Then R is $[Q^+]$-generated by any countable dense subring. If every such ring is $[S]$-generated then $[Q^+] < [S]$.

Proof Similar to that of Theorem 5.7 above.

Conjecture 5.9 The above result is true for the class of $\mathcal{P}_{22}$-metrizable topological rings.

Theorem 5.10 Let $(F, \mathcal{T})$ be an $\mathcal{P}_{31}$-metrizable topological field, and let $\mathcal{S} \subset \mathcal{H}$ be defined as in Theorem 4.6. Then $(F, \mathcal{T})$ is $[Q^+, V_{31}, \mathcal{S}]$-generated if and only if F is isomorphic to a subfield of R.

Proof (sufficiency) If K is a subfield of R, $j: F \to K$ an isomorphism, and $f: F \to R$ generates the topology $\mathcal{T}$ and satisfies $\mathcal{P}_{31}$, then there exists an a in $(0,1]$ such that

$$f(x) = |j(x)|^a$$

by Ostrowski (Bourbaki [9, Theorem 2, page 410]). Then $Q \subset K$. Let $H = j^{-1}(Q)$ and $g = f^{1/a}$. Then H is dense in F, g satisfies $\mathcal{P}_{31}$, $\mathcal{T} = \mathcal{T}_f = \mathcal{T}_g$ and $g(H) \subset \mathbb{Q}^+$.

(necessity) Suppose the result is false. Regard F as a subfield of $\mathbb{C}$ and suppose that there exists a subfield $H \subset F$ such that H is dense in F and $|x| \in \mathbb{Q}$ for all x in H. Then, because F is not contained in $\mathbb{R}$, there exists a number $c + ib$ in H with $b \neq 0$. Because $\mathbb{Q} \subset H$

$$\{((x - c)^2 + b^2)^{\frac{1}{2}} : x \in \mathbb{Q}\} \subset \mathbb{Q} .$$

Let $x = 0$. This implies $c^2 + b^2$ is in $\mathbb{Q}$. Let $x = 1$. Then $1 - 2c + c^2 + b^2 \in \mathbb{Q}$ and so $c \in \mathbb{Q}$. Let $x = c$. This implies $(b^2)^{\frac{1}{2}} \in \mathbb{Q}$ and thus $b \in \mathbb{Q}$. Finally if we let $x = c + b$ then $(2b^2)^{\frac{1}{2}}$ is in $\mathbb{Q}$ and therefore, because $b \neq 0$, $2^{\frac{1}{2}}$ is in $\mathbb{Q}$. This contradiction shows that $|x|$ is not in $\mathbb{Q}$ for at least one x in H and so F is not $[\mathbb{Q}^+, V_{31}, \mathcal{S}]$-generated.

We will consider subspaces of topological vector spaces in a similar manner. The symbol $\mathbb{K}$ means a field belonging to the restricted family of fields introduced in the last section of Chapter IV above.

Definition 5.11 We say that a $\mathbb{K}$ TVS E is dispersed or dispersed over $\mathbb{K}$ if the image, under the embedding

$$j: E \to \hat{E}$$

of E into its completion, of any $\mathbb{K}$ linear independent set is $\hat{\mathbb{K}}$ linear indepedent, where $\hat{\mathbb{K}}$ is the completion of $\mathbb{K}$. If $\mathbb{K}$ is complete then any $\mathbb{K}$ TVS is dispersed. The $\mathbb{Q}$ TVS $\mathbb{Q} + 2^{\frac{1}{2}}\mathbb{Q}$ with the topology inherited from $\mathbb{R}$ is not dispersed whereas $\mathbb{Q} + i\mathbb{Q}$ with the topology inherited from $\mathbb{C}$ is dispersed.

Theorem 5.12 Let E be a dispersed TVS over a ρ_{31}-metrizable topological field $\mathbb{K}$. Suppose that $\dim_{\mathbb{K}} E = n < \infty$. Then E is isomorphic to $\mathbb{K}^n$ algebraically and topologically.

Proof If $\mathbb{K}$ is complete the result is well known. If not let $\{e_1,\ldots,e_n\}$ be a basis for E over $\mathbb{K}$. We will prove that $\{j(e_1),\ldots,j(e_n)\}$ is a basis for $\hat{E}$ over $\hat{\mathbb{K}}$: it is linearly independent because E is dispersed. The space spanned by $\{j(e_1),\ldots,j(e_n)\}$ is closed (because it is complete) and contains $j(E)$ which is dense in $\hat{E}$. Thus the set of vectors spans $\hat{E}$. Therefore $\dim_{\hat{\mathbb{K}}}\hat{E} = n$ also. Let $f: (\hat{\mathbb{K}})^n \to E$ be defined by

$$(a_i)_{1 \leq i \leq n} \to a_1 j(e_1) + \ldots + a_n j(e_n) .$$

Then f is an isomorphism and its restriction to $\mathbb{K}^n \subset (\hat{\mathbb{K}})^n$ is an isomorphism onto $j(E)$ which is isomorphic with E. This completes the proof.

Theorem 5.13 Let $(E, \mathcal{T})$ be a $\mathcal{P}_{41}$-metrizable TVS over $\mathbb{R}$ with $\dim_{\mathbb{R}} E \leq \aleph_0$. Then E is $[\mathbb{Q}^+, V_{\mathbb{R}}, \mathcal{H}]$-generated.

Proof Assume $E \neq \{0\}$. Let $\{e_i : i \in J\}$ be a basis for E over $\mathbb{R}$. Then the subset

$$W = \{q_1 e_{i_1} + \ldots + q_n e_{i_n} : n \in \mathbb{N} , q_j \in \mathbb{Q} \text{ for } 1 \leq j \leq n\}$$

is a countable dense $\mathbb{Q}$ vector subspace of E. Let the function $f: E \to \mathbb{R}$ satisfy $\mathcal{P}_{41}$ and $\mathcal{T} = \mathcal{T}_f$. Then $f(W)$ is denumerable. Let $g \in \hom(\mathbb{R}^+, \mathbb{R}^+)$ be such that $g(f(W)) \subset \mathbb{Q}^+$. Then $h = g \circ f : E \to \mathbb{R}$ satisfies $\mathcal{P}_{41}$, $\mathcal{T} = \mathcal{T}_h$ and $h(W) \subset \mathbb{Q}^+$. This completes the proof.

Metric and Normed Spaces

We have seen above that, for S and T in $\mathcal{H}$, $[S] = \{S\}$ and $S < T$ if and only if $S \subset T$ when we are in the category of metric spaces, by Theorem 1.16. Instead of "[S]-generated" we will write "S-generated". It is not true to say that every separable metric space is $\mathbb{Q}^+$-generated: to see this consider the example $\mathbb{Q}$ with the metric $d(x,y) = \pi |x-y|$. In the sequel we will exhibit spaces with and without nice $\mathbb{Q}^+$-metric subspaces and present an example to show how this type of investigation is related to Diolphantine analysis.

<u>Example 5.14</u> Let $X = C[0,1]$, the set of continuous real valued functions on I and, for each f in X, let

$$\| f \| = \sup \{f(x): x \in I\}$$

be the supremum norm on X. Let M be the family of all functions f in X such that there exists rational numbers $0 = x_o < x_1 < \ldots < x_n = 1$ with f linear on $[x_{i-1}, x_i]$ for $1 \leq i \leq n$ and $f(x_i) \in \mathbb{Q}$ for $0 \leq i \leq n$. Then $M \subset X$ is a countable and dense $\mathbb{Q}$ linear subspace and $\| f \| \in \mathbb{Q}$ for each f in M. Thus X is $\mathbb{Q}^+$-generated as a real normed vector space.

<u>Example 5.15</u> We will consider Euclidean spaces $(R^n, \|\cdot\|_q)$ where n is in $\mathbb{N}$, q is in $\mathbb{N} \cup \{\infty\}$ and the norm $\| x \|_q = [\,|x_1|^q + \ldots + |x_n|^q]^{1/q}$ if $q < \infty$ and $\| x \|_\infty = \max \{|x_i| : 1 \leq i \leq n\}$ where $x = (x_1, \ldots, x_n) \in R^n$. We are interested in finding

pairs (n,q) for which $(\mathbb{R}^n, \|\cdot\|_q)$ is $\mathbb{Q}^+$-generated as a vector space or metric space.

Theorem 5.16 Let q be 1 or ∞. Then for all n in $\mathbb{N}$ $(\mathbb{R}^n, \|\cdot\|_q)$ is $\mathbb{Q}^+$-generated by the $\mathbb{Q}$-linear subspace $\mathbb{Q}^n$.

Note that when q is not in $\{1,\infty\}$ there exists an x in $\mathbb{Q}^n$ such that $\| x \|_q \notin \mathbb{Q}$ when $n \geq 2$.

Let $M(n,q) = \{x \in \mathbb{R}^n : \| x \|_q \in \mathbb{Q}\}$. We note that $M(n,$ is a subgroup of $\mathbb{R}^n$ if and only if $n = 1$.

Theorem 5.17 Let $n \in \mathbb{N}$. Then $M(n,2) \cap \mathbb{Q}^n$ is dense in

Proof If $n = 1$ then $M(1,2) \cap \mathbb{Q} = \mathbb{Q}$ and the result is certainly true. If $n \geq 2$ we will prove the result assuming that $M(n,2) \cap \mathbb{Q}^n \cap S^{n-1}$ is dense in the $n-1$ sphere S^{n-1}. If this is so let $x \in \mathbb{R}^n - \{0\}$ and let $\varepsilon > 0$ be given with $0 < \varepsilon < \| x \|_2$. Then $x / \| x \|_2 \in S^{n-1}$ and there exists a point y in $M(n,2) \cap \mathbb{Q}^n \cap S^{n-1}$ with $\| y - x / \| x \|_2 \|_2 < 1/4\ \varepsilon$. Let b in $\mathbb{Q}$ be such that $| b - \| x \|_2 | < 1/4\ \varepsilon$. Then $b.y$ is in $M(n,2) \cap \mathbb{Q}^n$ and

$$\begin{aligned} \| b.y - x \|_2 &= \| b.y - \| x \|_2 . y + \| x \|_2 . y - x \|_2 \\ &\leq | b - \| x \|_2 | .1 + \| x \|_2 . \| y - x / \| x \| \\ &< 1/4\ \varepsilon + 1/4 \| x \|_2 . \varepsilon \ . \end{aligned}$$

If $x = 0$ and y is any point in $\mathbb{Q}^n \cap S^{n-1}$ and m in $\mathbb{N}$ satisfies $1/m < \epsilon$ then $\| x - y/m \|_2 < \epsilon$.

It remains to prove that $\mathbb{Q}^n \cap S^{n-1}$ is dense in S^{n-1} for all $n \geq 2$. Let $n = 2$ and let $(u,v) \in \mathbb{Q}^2$. Then

$$\{((u^2-v^2)/(u^2+v^2),\ 2uv/(u^2+v^2)) : (u,v) \in \mathbb{Q}^2\}$$

is a dense subset of S^1 . (To see that this is so let $u = \cos(a/2)$ and $v = \sin(a/2)$.) We will proceed using induction.

If $n > 1$ assume that $\mathbb{Q}^k \cap S^{k-1}$ is dense in S^{k-1} for all $k \leq n$ and consider $\mathbb{Q}^{n+1} \cap S^n$. Let $\epsilon > 0$ be given and let $x \in S^n$, $x = (x_1,\dots,x_{n+1})$ where $x_1^2 + \dots + x_n^2 = 1$. If $x_{n+1} = 1$ or -1 we are through. If not choose y in $\mathbb{R}$ so that $x_{n+1}^2 + y^2 = 1$ and (s,t) in $S^1 \cap \mathbb{Q}^2$ so that $\|(x_{n+1}, y) - (s,t)\|_2 < \epsilon$ by the case $n = 2$ above. Now $(x_1/y,\dots,x_n/y) \in S^{n-1}$ because $y \neq 0$, and so, by the inductive assumption, there exists a point $(a_1,\dots,a_n)$ in S^{n-1} such that $\|(a_1,\dots,a_n) - (x_1/y,\dots,x_n/y)\|_2 < \epsilon$. Then $(a_1.t,\dots,a_n.t,\ s)$ is in $S^n \cap \mathbb{Q}^{n+1}$ and

$$\|(a_1.t,\dots,a_n.t,\ s) - (x_1,\dots,x_n,x_{n+1})\|_2^2$$

$$\leq \|(a_1.t,\dots,a_n.t) - (x_1,\dots,x_n)\|_2^2 + |s - x_{n+1}|^2$$

$$\leq \{|t| . \|(a_1,\dots,a_n) - (x_1/y,\dots,x_n/y)\|_2^2 + |(t/y) - 1| \times$$

$$\times \|(x_1,\dots,x_n)\|_2\}^2 + |s - x_{n+1}|^2$$

$< \{1.\epsilon + \epsilon.1\}^2 + \epsilon^2 = 5\epsilon^2$ and the proof is complete.

<u>Remark</u> Fermat's Last Theorem is equivalent to the conjecture $\mathbb{Q}^2 \cap M(2,q) = \mathbb{Q} \times \{0\} \cup \{0\} \times \mathbb{Q}$ for all $q > 2$.

<u>Conjecture 5.18</u> There does not exist a countable dense subset $X \subset \mathbb{R}^n$ such that $\| x - y \|_2 \in \mathbb{Q}$ for all x and y in X when $n > 1$.

<u>Example 5.19</u> <u>Metrizable Polyhedra</u> Let K be a simplicial complex whose related space $|K|$ is metrizable. If α and β are in $|K|$ let

$d(\alpha,\beta) = \sum_{v \in K_o} |\alpha(v) - \beta(v)|$ where K_o is the set of vertices of K. Then d is a metric compatible with the topology of $|K|$. Let $P = \{\alpha \in |K| : \alpha(v) \in \mathbb{Q}$ for each v in $K_o\}$. Then if α and β are in P, $d(\alpha,\beta)$ is in $\mathbb{Q}$. Let $sd^n K$ be the iterated barycentric subdivisions of K and consider $(sd^n K)_o$ to be a subset of $|K|$. Let $V = \bigcup_{n=0}^{\infty} (sd^n K)_o$ be the union of the vertices. Then $V \subset P \subset |K|$, V is dense in $|K|$ and forms a $\mathbb{Q}^+$-generating subspace of $|K|$.

Completions of compatible metrics on $\mathbb{Q}$

Let $(\mathbb{Q},\mathcal{T})$ be the rational numbers with their usual topology. If d is a compatible metric on $\mathbb{Q}$ then the completion $(\hat{\mathbb{Q}},\hat{d})$ is a complete separable metric space without isolated points. The following theorem indicates that all complete separable metric spaces without isolated points can be generated in this manner.

Theorem 5.20 (X,h) be a complete separable metric space without isolated points. Then there exists a metric d on the rational numbers $\mathbb{Q}$ generating the usual topology $\mathcal{T}$ and satisfying $(\hat{\mathbb{Q}},\hat{d})$ and (X,h) are isometric.

Proof Let $D \subset X$ be countable and dense. Because X does not have isolated points D is dense in itself and therefore homeomorphic to the rational numbers, Sierpinski [59, page 107]. Let $f\colon \mathbb{Q} \to D$ be a homeomorphism. Then

$$d(x,y) = h(f(x), f(y))$$

is a compatible metric on $\mathbb{Q}$. Finally because $f(\mathbb{Q}) = D$, D is dense in X, X is complete and $f\colon (\mathbb{Q},d) \to (X,h)$ is an isometry it follows that (X,h) is the completion of $(\mathbb{Q},d)$.

In the category $\mathcal{U}$ of metrizable uniform spaces we can find a metric d on $\mathbb{Q}$ which takes values in $\mathbb{Q}^+$. This is not true in general for the category $\mathcal{M}$ of metric spaces and isometries. (Consider a suitable metric space whose metric takes values in $\{\pi/n: n \in \mathbb{N}\}$.)

<u>Theorem 5.21</u> Let $\mathcal{T}$ be the usual topology on $\mathbb{Q}$. Let $(X,\mathfrak{U})$ be a complete metrizable uniform space whose induced topology is separable and has no isolated points. Then there exists a metric d on $\mathbb{Q}$, generating $\mathcal{T}$, having its range in $\mathbb{Q}^+$, and satisfying $(\hat{\mathbb{Q}},\hat{\mathfrak{U}}_d)$ and $(X,\mathfrak{U})$ are uniformly isomorphic, where $\mathfrak{U}_d$ is the uniform structure on $\mathbb{Q}$ induced by d.

<u>Proof</u> Let h be a metric on X satisfying $\mathfrak{U} = \mathfrak{U}_h$ let D be a countable dense subset and let $f: \mathbb{Q} \to D$ be a homeomorphism. Then, by the proof of Theorem 1.30, there exists a function g in $\hom(R^+,R^+)$ satisfying $g(h(D^2)) \subset \mathbb{Q}^+$ and $\mathfrak{U}_h = \mathfrak{U}_{g\circ h}$ by Theorem 1.8. Let

$$d(x,y) = g(h(f(x), f(y))$$

when x and y are in $\mathbb{Q}$. Then $(X,\mathfrak{U})$ is (uniformly isomorphic to) the completion of $(\mathbb{Q},\mathfrak{U}_d)$.

In th light of these last two results we may take the point of view that instead of being "given" a complete metric space we are "given" a copy of Q and a compatible metric d. The pair (Q,d) may be easier to specify for some purposes than the original complete metric space. We could regard the pair (Q,d) as a concrete object and its completion as abstract.

Let $R(Q)$ be the family of compatible metrics on Q (studied in [14] and for arbitrary metric spaces in Vaughn [66]). Then the natural operations on $R(Q)$ will induce operations on the set of complete separable metric spaces when the spaces are specified as described in the previous paragraph. For example define $(\hat{Q},\hat{d}) + (\hat{Q},\hat{h})$ to be $(\hat{Q}, \widehat{d+h})$ and $a(\hat{Q},\hat{d})$ to be $(\hat{Q}, \widehat{a.d})$ when $a > 0$ and d and h are in $R(Q)$.

We note that (Q,d) cannot be recovered from $(\hat{Q},\hat{d})$. In special circumstances the metric d is determined by $(\hat{Q},\hat{d})$ up to a uniform isomorphism as outlined in the following paragraph.

Let (X,d) be a metric space. We say (X,d) satisfies property q if whenever $A \subset X$ and $B \subset X$ are countable and dense there exists a uniform isomorphism $f: X \to X$ satisfying $f(A) = B$. If (X,d) is separable and complete and satisfies property q then the metric h on Q satisfying $(\hat{Q},\hat{h}) = (X,d)$ is well determined up to a uniform isomorphism, that is to say, if k is another metric on Q satisfying $(\hat{Q},\hat{k})$ is isometric to $(\hat{Q},\hat{h})$ then there exists a uniform isomorphism $g\colon (Q,h) \to (Q,k)$. The Hilbert cube satisfies property q , Fort [26], as does any compact manifold without boundary, Bennett [5]. The unit interval does not satisfy property q : if $A = Q \cap [0,1]$ and $B = A - \{0,1\}$, then A and B are countable and dense but no homeomorphism f of $[0,1]$ satisfies $f(A) = B$.

Let (X,d) be a complete separable metric space without isolated points. The essence of Theorem 5.20 is the existence of a topological embedding $e\colon Q \to X$ with $cl(e(Q)) = X$. Then the topological space $(Q,\mathcal{T})$ satisfies the following universal property: let $(P,\mathfrak{S})$ be a topological space with the property that for all metrizable separable topological spaces X without isolated points there exists an embedding $e\colon P \to X$ with $cl(e(P)) = X$, then necessarily P is homeomorphic to Q .

I conjecture that Q satisfies the stronger universal property:

Conjecture 5.23 Let $(P,\mathfrak{S})$ be a topological space having the property that for all complete separable metric spaces (X,d) without isolated points there exists a

topological embedding $e: P \to X$ with $cl(e(P)) = X$.
Then P is homeomorphic to $\mathbb{Q}$.

Theorem 5.24 Let (x_n) be a sequence in $\mathbb{Q}$ which is d-Cauchy for all metrics d in $R(\mathbb{Q})$. Then there exists a point x in $\mathbb{Q}$ such that x_n converges to x .

Proof If the result of the theorem were not true the sequence, because it is caucy with respect to the usual metric on $\mathbb{Q}$, will converge to an irrational number y in $\mathbb{R}$. We may assume that the given sequence is monotonically increasing and that its values are distinct. Let (y_n) be a sequence of irrational numbers satisfying

$$y_1 < x_1 < y_2 \dots y_n < x_n < y_{n+1} \dots < y .$$

Define a homeomorphism $f: \mathbb{Q} \to \mathbb{Q}$ as follows:

$(y,\infty) \cap \mathbb{Q}$ is mapped homeomorphically onto $(-\infty,\pi) \cap \mathbb{Q}$, $(-\infty,y_1) \cap \mathbb{Q}$ onto $(-\pi,\pi)$, and for each j in $\mathbb{N}$, (y_j, y_{j+1}) is mapped onto $(\pi+j-1, \pi+j) \cap \mathbb{Q}$. Define a compatible metric d on $\mathbb{Q}$ by the rule

$$d(x,y) = | f(x) - f(y) | .$$

Then $d(x_n, x_{n+2}) \geq 1$ for all n in $\mathbb{N}$, and therefore $(x_n)_{n\in \mathbb{N}}$ is not d-cauchy. This completes the proof.

Conjecture 5.25 Let $(X, \mathcal{T})$ be a metrizable topological space and let $R(X)$ be the set of metrics on X compatible with the topology $\mathcal{T}$. Then a sequence $(x_n)_{n \in \mathbb{N}}$ in X is convergent in X, if and only if, (x_n) is d-cauchy for all d in $R(X)$.

Example 5.26 The number e is irrational: Define a function $f: \mathbb{Q} \to \{0,1\}$ by setting

$$f(x) = \begin{cases} 1 & \text{if } (1 + 1/n)^n \leq x \text{ for all } n \in \mathbb{N}, \\ 0 & \text{otherwise.} \end{cases}$$

Let $d(x,y) = |x - y| + |f(x) - f(y)|$ for x and y in $\mathbb{Q}$. The map d is a compatible metric on $\mathbb{Q}$. Let $a_{2m} = (1 + 1/m)^m$ and $a_{2m+1} = (1 + 1/m)^{m+1}$ for m in $\mathbb{N}$. Then $\lim_{n \to \infty} a_n = e$ in R, but, because $d(a_{2m}, a_{2m+1}) \geq 1$ for all m, the sequence $(a_n)_{n \in \mathbb{N}}$ is not d-cauchy. Therefore, by Theorem 5.24, e is not in $\mathbb{Q}$.

We will prove a fixed point theorem (yet another!) modelled on the contraction principle of Peano [46], Picard [48], Banach [2], Cacciopoli [17] and others.

Definition 5.27 Let X be a set and $f: X \to X$ a function. Let M be a family of metrics on X. We say that f is an M-contraction on X if for all d in M there exists an r in the interval $[0,1)$, depending on d , such that

$$d(f(x), f(y)) \leq r.d(x,y)$$

for all x and y in X .

<u>Theorem 5.28</u> Let $f: \mathbb{Q} \to \mathbb{Q}$ be an $R(\mathbb{Q})$-contraction. Then f has a unique fixed point in $\mathbb{Q}$.

<u>Proof</u> Let x be any point in $\mathbb{Q}$. The proof of the contraction mapping theorem in a complete metric space shows that the sequence of iterates $(f^n(x))_{n \in \mathbb{N}}$ is d-cauchy for all d in $R(\mathbb{Q})$. Therefore this sequence converges in $\mathbb{Q}$ by Theorem 5.24. The limit of the sequence is the unique fixed point of f .

Apart from constant functions, $R(\mathbb{Q})$-contractions may be difficult to find. On the other hand one does not expect to find too many fixed point theorems for continuous functions on $\mathbb{Q}$.

<u>The extension problem for S-metrics</u>

We will show my means of examples that S-metrics frequently do not have S-metric extensions unless [S] = [I], in which case we have the fine extension theorem of R.H. Bing [6] .

Example 5.29 Let (X,d) be any complete connected separable metric space without isolated points and let T be a subset of X homeomorphic to the Cantor ternary set. Then no compatible S metric on T has an extension to a compatible S metric on X if S is not a neighborhood of zero. The set T is a closed subset of X.

Example 5.30 Let the rational numbers be given their usual metric and let X be an extension of $\mathbb{Q}$ by one more point a. If X is to contain a as an isolated point then we may define a $\mathbb{Q}$-metric d on X as follows:

$$d(a,x) = 1 + |2 - x| \quad , \text{ for all } x \text{ in } \mathbb{Q} ,$$

$$d(a,a) = 0 ,$$

$$d(x,y) = |x - y| \quad \text{when } x \text{ and } y \text{ are in } \mathbb{Q} .$$

The metric d is then a $\mathbb{Q}$-metric extension of the usual metric.

If X is to contain $\mathbb{Q}$ as a dense subset however, we will show that the usual metric on $\mathbb{Q}$ does not have a $\mathbb{Q}$-metric extension to a compatible metric on X. Suppose that such an extenstion exists and let (X,d) be the resulting space. Then there exists a sequence $(r_j)_{j \in \mathbb{N}}$ of rational numbers converging to a in X. Because the sequence is d-convergent it is d-cauchy and therefore cauchy in the usual metric on $\mathbb{Q}$. Then there exists a real number b such that $|r_j - b| \to 0$. The number b must be irrational because a does not belong to $\mathbb{Q}$ and limits are unique in X .

Because d satisfies the triangle law for metrics we have:

$$|d(x,z) - d(z,y)| \leq d(x,y) \leq d(x,z) + d(z,y)$$

for all x, y and z in X . Then, for all j in $\mathbb{N}$ we have

$$|d(a,r_j) - |r_j - x|| \leq d(a,x) \leq d(a,r_j) + |r_j - x| .$$

Letting j tend to infinity we obtain in the limit

$$d(a,x) = |b - x|$$

for all x in $\mathbb{Q}$. In particular d is not a $\mathbb{Q}$-metric on X .

The reader will note that we can easily produce the desired topology on X if we change the metric on $\mathbb{Q}$.

The previous example (5.30 second part) has an easy application to the following theorem:

<u>Theorem 5.31</u> Let $\mathbb{Q}$ have the usual metric and let (X,d) be a $\mathbb{Q}$-metric space. Let

$$i: \mathbb{Q} \to X$$

be an isometry (not necessarily surjective). Then $i(\mathbb{Q})$ is closed in X . If (X,d) is a complete $\mathbb{Q}$-metric space then no such isometries exist.

Embedding Metric Spaces in Banach Algebras

Definition 5.32 Let (X,d) be a metric space and let q be a real number with $1 \leq q$. We say d is a q-metric if, for all x, y and z ,

$$d(x,y)^q \leq d(x,z)^q + d(z,y)^q \ .$$

If this is so we say (X,d) is a q-metric space. We say d is an ultra-metric if, for all x, y and z ,

$$d(x,y) \leq \max\{d(x,z),\ d(z,y)\} \ .$$

Then d is an ultra-metric if and only if it is a q-metric for all $q \geq 1$.

Let B be a Banach algebra over the real or complex field. We say B is a q-Banach algebra if the metric induced by the norm on B is a q-metric.

Theorem 5.33 Let $q \geq 1$ and let (X,d) be a q-metric space. Then there exists a q-Banach algebra B over the real numbers and an isometry

$$i: X \to B$$

mapping X into B.

<u>Proof</u> We will use a method modelled on that of Kuratowski [33]. Let B be the set of real valued continuous and bounded functions on X, let x_o be a point in X, and for each x and y in X let

$$f_x(y) = d(x,y)^q - d(y,x_o)^q .$$

The set B becomes a q-Banach algebra when we define the norm

$$\| f \| = \sup\{ | f(x) |^{1/q} : x \in X\}$$

for each f in B. Define a map $i: X \to B$ by $x \to f_x$. Then

$$| f_x(y) - f_z(y) | = | d(x,y)^q - d(z,y)^q | \leq d(x,z)^q$$

because d is a q-metric. Thus $\| f_x - f_z \| \leq d(x,z)$. Also $| f_x(z) - f_z(z) | = d(x,z)^q$ showing that $\| f_x - f_z \| = d(x,z)$ for all x and z in X. Therefore i is an isometry.

When $q = \infty$ we must replace the field of scalars R by a p-adic field Q_p. (One reason: if B is a Banach algebra over R and $\| f+g \| \leq \max\{\| f \|, \| g \|\}$ for all

f and g in B then $B = \{0\}$.) These so-called non-archimedean Banach algebras are well known; see for example [8], [4], [44], [51], and [58].

Let p be an odd prime and let $H_p = \{p^n : n \in \mathbb{Z}\} \cup \{0\}$. If (X,d) is an H_p-metric space it is easy to see that necessarily $d(x,y) \leq \max\{d(x,z), d(z,y)\}$ for all x, y and z in X.

<u>Conjecture 5.34</u> Let (X,d) be a H_p-metric space. Then there exists a non-archimedean Banach algebra B over the field $\mathbb{Q}_p$ and an isometry $i: X \to B$.

Consider the class of complete separable metric spaces without isolated points. By an earlier theorem this class may be derived from the family of metric spaces

$$\{(\mathbb{Q},d) : d \in R(\mathbb{Q})\}$$

by taking completion. This means that all the metric and derived information about a space $(\hat{\mathbb{Q}},\hat{d})$ will already be carried by the original metric d on $\mathbb{Q}$. One may wish to replace $\mathbb{Q}$ by a homeomorphic copy. The metric values could be chosen to lie in the rationals, the dyadic rationals, the algebraic numbers, the computable numbers or any other countable and dense subset of $\mathbb{R}$ without causing any change in the completion when viewed as a uniform space.

The central idea is to relate properties of completions

$$(\hat{Q},\hat{d}) \ , \ (\hat{Q},\mathcal{T}_{\hat{d}}) \ , \ (\hat{Q},\hat{u}_{\hat{d}}) \ ,$$

and so on, with properties of the metric d. This has been done already for many traditional concepts. For instance the completion $(\hat{Q},\hat{d})$ is compact if and only if the metric d is totally bounded.

When developed in this manner the main problem is not to classify complete separable metric spaces up to homeomorphism, but to isolate the complete separable metric spaces which are completions of reasonable metrics on reasonable homeomorphs of Q. The criteria for reasonableness will depend on the requirements of the reasoner. The non unicity of the metric can be thought of as an advantage in that different metrics may be better for different purposes. Special metrics become important -- indeed we have the concept of the "usual metric" for many spaces.

This digitalization of a piece of analysis could form part of a bridge between scientific practice (measurement, computation, model construction, simulation, etc.) and the mathematically well developed, but often non constructive, theories of continuously varying quantities.

Here, after first developing a notion of connectedness for uniform spaces, we will study metric connectedness. It transpires that the rational numbers are a connected subset of the real numbers!

Uniform Connectedness

The theory is parallel to the theory of connectedness in topological spaces except for the existence of relationships between the connectedness properties of a uniform space and the connectedness properties of its completion.

Definition 5.35 Let $(X,\mathfrak{U})$ be a uniform space and let S be a non empty subset of X. We say S is disconnected or uniformly disconnected or uni-disconnected if there exists a U in $\mathfrak{U}$ and non empty subsets A and B in X satisfying

$$A \cup B = S \quad \text{and} \quad U(A) \cap U(B) = \emptyset .$$

We say S is connected or uniformly connected or uni-connected if it is not disconnected.

Example 5.36 If Q has its usual uniform structure then $Q \cap [0,1]$ is a connected subset of Q and $Q - [0,1]$ a disconnected subset of Q.

<u>Example 5.37</u> If $\mathbb{R}$ has its usual uniform structure then $\mathbb{R} - \{0\}$ is a connected subset.

<u>Theorem 5.38</u> Let $(X,\mathfrak{U})$ be a uniform space, S a non empty subset of X which is connected, and Y a subset of X satisfying

$$S \subset Y \subset cl(S)$$

where closure is taken in $\mathcal{T}_{\mathfrak{U}}$. Then Y is a connected subset. In particular the closure of a connected set is connected.

<u>Proof</u> Let Y be disconnected. Then there exist non empty subsets A and B in X and an entourage U in $\mathfrak{U}$ satisfying $A \cup B = Y$ and $U(A) \cap U(B) = \emptyset$. Because S is connected, $S \subset A$ or $S \subset B$. If $S \subset A$ there exists a point b in $B \cap Y$ and then $U(\{b\}) \cap A = \emptyset$, contradicting the fact that b is in $cl(S)$. Thus Y is connected.

<u>Theorem 5.39</u> Let $(X,\mathfrak{U})$ be a uniform space and let $(P_i)_{i \in I}$ be a family of non empty connected subsets with the property $P_i \cap P_j \neq \emptyset$ for all i and j in I. Then $P = \bigcup_{i \in I} P_i$ is a connected subset of X.

Proof Suppose P is disconnected. Then there exist non empty sets A and B in X and a U in $\mathfrak{U}$ with $A \cup B = P$ and $U(A) \cap U(B) = \emptyset$. If i is in I then $P_i \subset A$ or $P_i \subset B$ because P_i is connected. There exist i and j in I with $P_i \subset A$ and $P_j \subset B$ because A and B are not empty and their union is P. But $P_i \cap P_j$ is not empty giving a contradiction. This completes the proof.

Let $(X,\mathfrak{U})$ be a uniform space and let $x \in X$. By Theorem 5.39 the union of the connected subsets of X which contain x is connected. By Theorem 5.38 this set is closed. We call the set a connected component and represent it with the symbol $U(x)$. If $U(x) \cap U(y) \neq \emptyset$ then $U(x) = U(y)$ and the set of distinct components forms a partition of X. The space $(X,\mathfrak{U})$ is connected if and only if there is an x in X such that $X = U(x)$.

Theorem 5.40 Let $(X,\mathfrak{U})$ and $(Y,\mathfrak{B})$ be uniform spaces and $f\colon X \to Y$ a uniformly continuous function. If S is a non empty connected subset of X then $f(S)$ is a connected subset of Y.

<u>Proof</u> Let $f(S)$ be disconnected. Then there exists a V in $\mathfrak{B}$ and non empty sets A and B in Y with $A \cup B = f(S)$ and $V(A) \cap V(B) = \emptyset$. Then $S \subset f^{-1}(A) \cup f^{-1}(B)$. Because f is uniformly continuous there exists a U in $\mathfrak{U}$ such that if $(x,y) \in U$ then $(f(x), f(y)) \in V$. Suppose that there is a point x in $U(f^{-1}(A)) \cap U(f^{-1}(B))$. Then there exists a point $f(a)$ in A and a point $f(b)$ in B such that $(f(x), f(b)) \in V$ and $(f(x), f(a)) \in V$. But this means that $f(x) \in V(A) \cap V(B)$, a contradiction. Therefore $U(f^{-1}(A)) \cap U(f^{-1}(B)) = \emptyset$ and therefore $U(S \cap f^{-1}(A)) \cap U(S \cap f^{-1}(B)) = \emptyset$. Clearly the sets $S \cap f^{-1}(A)$ and $S \cap f^{-1}(B)$ are both non empty, because A and B are non empty. This is a contradiction on the connectedness of S.

<u>Theorem 5.41</u> Let $(X,\mathfrak{U})$ be a uniform space. If $(X,\mathcal{T}_{\mathfrak{U}})$ is top-connected then $(X,\mathfrak{U})$ is uni-connected.

<u>Proof</u> Let $(X,\mathfrak{U})$ be uni-disconnected. Then there exist non empty A and B in X and U in $\mathfrak{U}$ with $A \cup B = X$ and $U(A) \cap U(B) = \emptyset$. There exists an open entourage V in U with $V \subset U$. Then $V(A) \cap V(B) = \emptyset$ and $V(A) \cup V(B) = X$. The sets $V(A)$ and $V(B)$ are each open and non empty and therefore the topological space $(X,\mathcal{T}_{\mathfrak{U}})$ is disconnected.

Example 5.42 Let $\mathbb{R}$ have the usual uniform structure and let $S \subset \mathbb{R}$ be a non empty subset. Then S is connected if and only if $cl(S)$ is an interval.

Example 5.43 (Intermediate Values) Let $(X,\mathfrak{U})$ be a connected uniform space and let $f: X \to \mathbb{R}$ be a uniformly continuous function. If z in $\mathbb{R}$ and a and b in X are such that

$$f(a) \leq z \leq f(b)$$

in $\mathbb{R}$ then there exists a sequence $(x_n)_{n \in \mathbb{N}}$ in X such that $f(x_n)$ tends to z when n tends to infinity.

Theorem 5.44 The uniform space $(X,\mathfrak{U})$ is connected if and only if its completion $(\hat{X},\hat{\mathfrak{U}})$ is connected.

Proof Let $(X,\mathfrak{U})$ be connected and let

$$i: X \to \hat{X}$$

be the canonical unimorphism. Then, by Theorem 5.40, $i(X)$ is a connected subset of $\hat{X}$, and then, by Theorem 5.38, $\hat{X} = cl(i(X))$ is connected.

Now let $(X,\mathfrak{U})$ be disconnected. Then there exist non empty subsets A and B in X and V in $\mathfrak{U}$ with $A \cup B = X$ and $V(A) \cap V(B) = \emptyset$. Recall that $\hat{X}$ may be taken to be the set of minimal cauchy filters on X and

$i(x) = N_x$, the filter of neighborhoods of x for each x in X , Robertson and Robertson [54]. Let $\hat{V} = \{(F_1, F_2) : \text{some } V\text{-small set is in } F_1 \cap F_2\}$ be the connector on $\hat{X}$ corresponding to V on X . Then

(i) $\hat{V}(i(A)) \cap \hat{V}(i(B)) = \emptyset$: If not there exists a minimal cauchy filter F on X and points a in A and b in B such that $(N_a, F) \in \hat{V}$ and $(N_b, F) \in \hat{V}$. This means that $N_a \cap F$ contains a V-small set G say, and $N_b \cap F$ contains a V-small set H . This implies that $G \subset V(a)$ and $H \subset V(b)$ and therefore

$$G \cap H \subset V(A) \cap V(B) .$$

But $G \cap H$ is non empty because G and H are in the filter F and thus $V(A) \cap V(B) \neq \emptyset$, a contradiction. This proves (i).

(ii) $cl(i(A)) \cup cl(i(B)) = cl(i(A) \cup i(B))$: this follows from a general property of the closure operator or may be proved directly.

(iii) $cl(i(A)) \subset \hat{V}(i(A))$ and $cl(i(B)) \subset \hat{V}(i(B))$: because in any uniform space $(Y,\mathfrak{B})$, if $D \subset Y$ then $cl(D) = \cap \{U(D): U \in \mathfrak{B}\}$.

We may now complete the proof. By (ii) $\hat{X} = cl(i(A)) \cup cl(i(B))$. By (i) and (iii) $cl(i(A)) \cap cl(i(B)) = \emptyset$. The sets $cl(i(A))$ and

$cl(i(B))$ are non empty. If $\hat{U}$ in $\hat{\mathfrak{U}}$ satisfies $\hat{U}^2 \subset \hat{V}$ then $cl(i(A)) \subset \hat{U}(cl(i(A))) \subset \hat{U}^2(i(A)) \subset \hat{V}(i(A)$, and the corresponding inclusions with B replacing A , imply

$$\hat{U}(cl(i(A))) \cap \hat{U}(cl(i(B))) = \emptyset .$$

This shows that $\hat{X}$ is disconnected. The proof is complete.

The concept uni-connectedness is related to a well known notion of connectedness in metric and uniform spaces, Sierpinski [59, section 47a, page 86] and Bourbaki [10, part I, page 204]. See also Levine [37] and Collins [18]. The following definition is taken from Bourbaki:

Let $(X, \mathfrak{U})$ be a uniform space and let V in $\mathfrak{U}$ be a symmetric entourage. A finite sequence $(x_i : 0 \leq i \leq n)$ of points of X is said to be a V-chain if for all i with $0 \leq i < n$, $x_{i+1} \in V(x_i)$. The points x_o and x_n are called the ends of the V-chain, and they are said to be joined by the V-chain.

<u>Theorem 5.45</u> The uniform space $(X, \mathfrak{U})$ is uni-connected if and only if, for all V in $\mathfrak{U}$ and points x and y in X , there exists a V-chain joining x and y .

<u>Proof</u> Let $(X, \mathfrak{U})$ be disconnected. Let U in $\mathfrak{U}$ and non empty subsets A and B in X be such that $A \cup B = X$ and $U(A) \cap U(B) = \emptyset$. Then if $a \in A$ and $b \in B$, no U-chain in X can join a and b.

Now suppose that there exist points x and y in X and a symmetric entourage V in $\mathfrak{U}$ such that no V-chain joins x and y. Let
$A = \{z \in X :$ there exists a V-chain joining x to $z\}$ and let $B = X - A$. Then A and B are non empty, $V(A) = A$, and if U in $\mathfrak{U}$ satisfies $U^2 \subset V$, $U(A) \cap U(B) = \emptyset$. Therefore $(X, \mathfrak{U})$ is disconnected. This completes the proof.

The following theorem is equivalent to Bourbaki [10, page 204, Proposition 6] and analagous to Sierpinski [59, page 87].

<u>Theorem 5.46</u> Let $(X, \mathfrak{U})$ be compact and uni-connected. Then $(X, \mathcal{T}_{\mathfrak{U}})$ is top-connected.

<u>Example 5.47</u> If the uniform space $(X, \mathfrak{U})$ is uni-connected and complete then $(X, \mathcal{T}_{\mathfrak{U}})$ need not, in general, be top-connected:

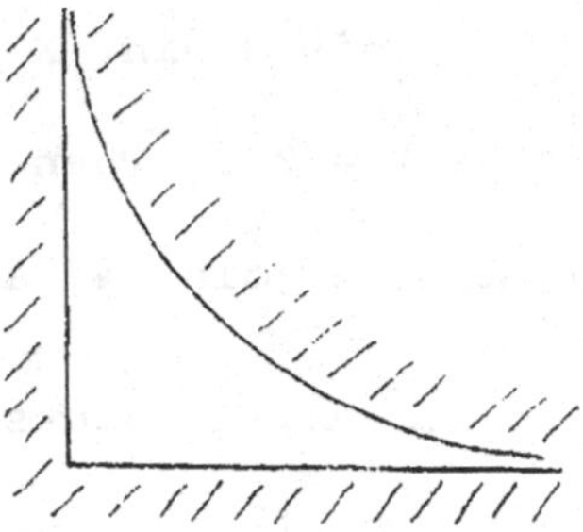

Figure 7

Let $X = \{(x,y): 0 < x,\ 0 < y, \text{ and } xy < 1\}$ and let $Y = R^2 - X$. Regard Y as a uniform subspace of R^2. Then Y is complete and uni-connected but not top-connected.

<u>Theorem 5.48</u> Let $(X,\mathcal{U})$ be connected and totally bounded. Then $(\hat{X},\mathcal{T}_{\hat{\mathcal{U}}})$ is connected.

<u>Proof</u> An easy corollary to Theorem 5.44 and Theorem 5.46.

We say the metric space (X,d) is connected or met-connected or d-connected if $(X,\mathcal{U}_d)$ is uni-connected. Then, by Theorem 5.44, the space (X,d) is d-connected if and only if its completion $(\hat{X},\hat{d})$ is $\hat{d}$-connected.

Computable Generation

We will define here the notion computable metric space and other related computable objects in a manner similar to that introduced by Rabin [52] for algebraic structures. Classical spaces such as spheres and tori are not computable but arise as the completions of computable subspaces (see the remarks following Conjecture 5.34). We will assume the notions recursive function, computable function and computable real number as known (see for example Rogers [55]).

Definition 5.49 Let $\mathbb{M} \subset \mathbb{R}$ be the set of computable numbers. We say a metric space (X,d) is computable if there exists a bijection $i: \mathbb{N} \to X$ and a computable function $f: \mathbb{N}^2 \to \mathbb{M}$ such that, for all $(n,m) \in \mathbb{N}^2$,

$$d(i(n), i(m)) = f(n,m) .$$

If (X,d) is computable and (Y,h) is an isometric metric space then (Y,h) is computable. If $d(x,y) = |x - y|$ then $(\mathbb{Q},d)$ is computable: let $i: \mathbb{N} \to \mathbb{Q}$ be a bijective recursive enumeration of $\mathbb{Q}$. Then $f(n,m) = |i(n) - i(m)|$ is computable and $\mathbb{Q} \subset \mathbb{M}$. Similarly if $n \in \mathbb{N}$ and $q \in \mathbb{N} \cup \{\infty\}$ then $(\mathbb{Q}^n, \|\cdot\|_q)$ is computable.

Definition 5.50 Let $(X,\mathcal{U})$ be a metrizable uniform space. We say $(X,\mathcal{U})$ is computable if there exists a metric d on X such that $\mathcal{U} = \mathcal{U}_d$ and (X,d) is a computable metric space.

Definition 5.51 Let $(X,\mathcal{T})$ be a metrizable topological space. We say $(X,\mathcal{T})$ is computable if there exists a metric d on X such that $\mathcal{T} = \mathcal{T}_d$ and (X,d) is a computable metric space.

Definition 5.52 Let $(G,\mathcal{T})$ be a metrizable topological group. We say $(G,\mathcal{T})$ is computable if there exists a bijection $i: \mathbb{N} \to G$ and a computable function $f: \mathbb{N}^2 \to \mathbb{M}$ such that

(i) if $h: G^2 \to G$ is the group operation, then

$$i^{-1} \circ h \circ (i \times i): \mathbb{N}^2 \to \mathbb{N}$$

is computable and

(ii) the function

$$d = f \circ (i^{-1} \times i^{-1}): G^2 \to \mathbb{M}$$

is a translation invariant metric on G generating the topology. Part (i) of this definition corresponds to Rabin [52, Definition 4, page 343].

For example the group $\mathbb{Z}$ with the discrete topology is a computable metrizable topological group.

Note that we could easily extend these notions to spaces which have richer (or poorer) algebraic structures.

Notation: if (X,d) is a metric space and $A \subset X$ a subset then d_A is the restriction of the metric d to the subset A.

Definition 5.53 We say (X,d) is computably generated if there exists a dense subset $A \subset X$ such that (A,d_A) is a computable metric space.

Definition 5.54 Let $(X,\mathcal{T})$ be a metrizable topological space. We say $(X,\mathcal{T})$ is computably generated if there exists a dense subset $A \subset X$ and a compatible metric d on X such that (A,d_A) is a computable metric space.

Definition 5.55 Let $(G,\mathcal{T})$ be a metrizable topological group. We say $(G,\mathcal{T})$ is computably generated if there exits a dense subgroup $A \subset G$ and a compatible metric d on G such that $(A,\mathcal{T}_{d_A})$ is a computable metrizable topological group.

For example R^n is a computably generated metrizable topological group. For all x and y in $\mathbb{C}$ let $d(x,y)$ be $|x - y|$. Then the circle group $(S^1,\mathcal{T}_d)$ is a computably generated metrizable topological group: let

$$A = \{((1-t^2)/(1+t^2),\ 2t/(1+t^2)) : t \in \mathbb{Q}\} \cup \{(-1,0)\} .$$

Then (A,d_A) is a computable metric group which is dense in S^1.

<u>Theorem 5.56</u> The finite product of computable topological spaces is computable.

<u>Proof</u> Let $(X,\mathcal{T})$ and $(Y,\mathfrak{S})$ be computable. Then there exist bijections $i: \mathbb{N} \to X$ and $j: \mathbb{N} \to Y$ and computable functions $f, g : \mathbb{N}^2 \to \mathbb{M}$ such that $f \circ (i^{-1} \times i^{-1}) = d$ is a metric on X generating $\mathcal{T}$ and $g \circ (j^{-1} \times j^{-1}) = h$ is a metric on Y generating the topology $\mathfrak{S}$. Let $(X \times Y, \mathcal{T} \times \mathfrak{S})$ be the topological product of X and Y, let $k: \mathbb{N} \to \mathbb{N}^2$ be a recursive enumeration and let

$$p((a,b),\ (x,y)) = d(a,x) + h(b,y)$$

be the metric on $X \times Y$ generating the product topology. Let $b = (i \times j) \circ k$. Then $b: \mathbb{N} \to X \times Y$ is a bijection. If $(s,t) \in \mathbb{N}^2$ then

$$p(b(s),\ b(t)) = f(k_1(s),k_1(t)) + g(k_2(s),k_2(t)) = a(s,t)$$

say, where $k_i: \mathbb{N} \to \mathbb{N}$ is the composition of k with the natural projection of $\mathbb{N}^2$ onto the ith factor for each i. Because k_i, f, g are computable and addition is computable in $\mathbb{M}$ the function $a: \mathbb{N}^2 \to \mathbb{M}$ is computable. This completes the proof.

<u>Example</u>: The torus $S^1 \times \ldots \times S^1$ ($n \in \mathbb{N}$ terms) is computably generated.

<u>Theorem 5.57</u> Let $d: \mathbb{Q}^2 \to \mathbb{Q}$ be a metric which is a computable function. Then $(\mathbb{Q},d)$ is a computable metric space.

<u>Final Remarks 5.58</u> We have shown (Theorem 5.20) that each separable complete metric space (X,d) without isolated points may be exhibited as the completion of $(\mathbb{Q},h)$ for some metric h on $\mathbb{Q}$ which is compatible with the usual topology and which takes values in $\mathbb{Q}$. When $(\mathbb{Q},h)$ is computable the corresponding completion (X,d) will be computably generated. In this way we have isolated a class of complete metric spaces which includes Euclidean spaces, the Cantor space and so on. Here we have merely defined the basic ideas.

Index of Notations

Structures of Spaces

$\mathfrak{A}$: arbitrary real-generated structure on a set, page 1

$\mathfrak{C}$: convergence structure on a set, page 15

$\mathfrak{M}$: metric structure on a set, page 10

$\mathfrak{N}$: proximity structure on a set, page 16

$\mathfrak{T}$: topological structure on a set, page 15

$\mathfrak{U}$: uniform structure on a set, page 6

Categories

$\mathcal{A}$ and $\mathcal{B}$: arbitrary real-generated categories, page 1

$\mathcal{C}$: metrizable convergence spaces and continuous functions, page 2

$\mathcal{D}$: topological spaces with the weak topology, page 101

$\mathcal{E}$: uniform spaces and uniformly continuous maps, page 3

$\mathcal{F}$: metrizable topological fields and continuous homomorphisms, page 111

$\mathcal{G}$: metrizable topological abelian groups, page 105

$\mathcal{J}$: clopen-paracompact metrizable topological spaces, page

$\mathcal{K}$: quasi uniform spaces, page 3

$\mathcal{L}$: metrizable topological rings, page 111

$\mathcal{M}$: metric spaces and isometries, page 2

$\mathcal{N}$: metrizable proximity spaces and p-continuous maps, page 2

$\mathcal{T}$: metrizable topological spaces and continuous functions, page 2

$\mathcal{U}$: metrizable uniform spaces, page 2

$\mathcal{V}$: metrizable general topological vector spaces, page 141

Subsets of the real numbers

$\frac{1}{2}D$ = $\{0,\frac{1}{2}\}$, page 135

D = $\{0,1\}$, page 8

D_m = $\{1/n_1 + \dots + 1/n_s$: for $1 \leq i \leq s \leq m$, n_i is a whole number$\} \cup \{0\}$, page 33

E = $\{1/3^n$: n is an integer$\} \cup \{0\}$, page 113

G = an arbitrary additive subgroup of the real numbers, page 58

H = $\{1/n$: n is a whole number$\} \cup \{0\}$, page 33

I = $[0,1]$, page 8

J_m = $\{I - D_m\} \cup \{0\}$ where m is a whole number, page 33

$\mathbb{M}$ = the computable real numbers, page 181

$\mathbb{N}$ = $\{1, 2, 3,\dots\}$, page 19

O = $\{0\}$, page 44

$\mathbb{P}$ = the irrational numbers, page 32

$\mathbb{P}^+$ = $\{\mathbb{P} \cap [0,\infty)\} \cup \{0\}$, page 32

$\mathbb{Q}$ = the rational numbers, page 32

$\mathbb{Q}^+$ = $\mathbb{Q} \cap [0,\infty)$, page 32

$\mathbb{R}$ = the real numbers, page 1

$\mathbb{R}^+$ = $\mathbb{R} \cap [0,\infty)$, page 33

S and T stand for arbitrary subsets of $\mathbb{R}$, page 2

W = $\{1/3^n$: n is a whole number$\} \cup \{0\}$, page 36

X_m = $\{1/2^{n_1} + \ldots + 1/2^{n_s}\}$: for $1 \leq i \leq s \leq m$, n_i is a whole number$\} \cup \{0\}$, page 91

$\mathbb{Z}$ = $\{0, \pm 1, \pm 2, \ldots\}$, page 56

$\mathbb{Z}^+$ = $\{0, 1, 2, \ldots\}$, page 56

Families of subsets of the real numbers

$\mathcal{H}$: subsets of the positive reals which include 0 , page 5

$\mathcal{J}$: non empty subsets of the real numbers, page 2

$\mathcal{O}$: neighborhoods of 0 in $\mathbb{R}^+$, page 41

$\mathcal{S}$: arbitrary non empty family of subsets, page 2

$\mathcal{W}$: positive parts of additive subgroups of $\mathbb{R}$, page 58

$\mathcal{X}$: closed subsets of the positive reals which include 0 , page 46

$\mathcal{Y}$: semimodules in $\mathbb{R}^+$, page 56

$\mathcal{Z}$: closed subgroups of $\mathbb{R}$, page 59

Properties of structure generating functions

Functions of two variables (page 93 and page 105)

1. $d(x,y) \geq 0$ for all x and y, $d(x,x) = 0$ for all x,
2. if $d(x,y) = 0$ then $x = y$,
3. $d(x,y) = d(y,x)$ for all x and y,
4. $d(x,y) \leq d(x,z) + d(z,y)$ for all x, y, and z,
5. $d(x,y) \leq \max\{d(x,z), d(z,y)\}$ for all x, y, and z,
6. $d(x+a, y+a) = d(x,y)$ for all x, y, and a.

$\mathcal{P}_1 = \{1,2,3,5\}$

$\mathcal{P}_2 = \{1,2,3,4\}$

$\mathcal{P}_3 = \{1,2,3\}$

$\mathcal{P}_4 = \{1,2\}$

$\mathcal{P}_5 = \{1,2,4\}$

$\mathcal{P}_6 = \{1,2,3,4,6\}$

$\mathcal{P}_{10} = \{1,3,5\}$

$\mathcal{P}_{20} = \{1,3,4\}$

$\mathcal{P}_{30} = \{1,3\}$

$\mathcal{P}_{40} = \{1\}$

$\mathcal{P}_{50} = \{1,4\}$

<u>Functions of one variable</u> (page 109 and page 141)

1. $f(x) = 0$ if and only if $x = 0$,
2. $f(x) = f(-x)$ for all x ,
3. $f(x+y) \leq f(x) + f(y)$ for all x and y ,
4. $f(x+y) \leq \max\{f(x), f(y)\}$ for all x and y ,
5. $f(x.y) \leq \min\{f(x), f(y)\}$ for all x and y ,
6. $f(x.y) \leq f(x).f(y)$ for all x and y ,
7. $f(x.y) = f(x).f(y)$ for all x and y ,
8. if $a \to 0$ in $\mathbb{K}$ and $x \in V$ then $f(ax) \to 0$ in $\mathbb{R}$,
9. if $a \in \mathbb{K}$ and $|a| \leq 1$ then $f(ax) \leq f(x)$ for all x

$\mathcal{P}_{21} = \{1,2,3,5\}$ $\qquad$ $\mathcal{P}_{22} = \{1,2,3,6\}$

$\mathcal{P}_{23} = \{1,2,4,5\}$ $\qquad$ $\mathcal{P}_{24} = \{1,2,4,6\}$

$\mathcal{P}_{31} = \{1,2,3,7\}$ $\qquad$ $\mathcal{P}_{32} = \{1,2,4,7\}$

$\mathcal{P}_{41} = \{1,2,3,8,9\}$

References

[1] R. Arens and J. Dugundji, "A remark on the concept of compactness", Portugaliae Math. 9 (1950), 141-143.

[2] S. Banach, "Sur les opérations dans les ensembles abstraits et leurs applications aux équations integrales", Fund. Math. 3 (1922), 133-181.

[3] B. Banaschewski, "Über nulldimensionale raüme", Math. Nachr. 13 (1955), 129-140.

[4] E. Beckenstein, "On regular non-archimedean Banach algebras", Arch. Math. 19 (1968), 423-427.

[5] R. Bennett, "Countable dense homogeneous spaces", Fund. Math. 74 (1972), 189-194.

[6] R.H. Bing, "Extending a metric", Duke Math. J. 14 (1947), 511-519.

[7] R.H. Bing , "Metrization of topological spaces", Canad. J. Math. 3 (1951), 175-186.

[8] W. Blum, "Über kommutative nichtarchimedische Banach algebren", Arch. Math. 24 (1973), S. 493-498.

[9] N. Bourbaki, <u>Commutative Algebra</u>, Addison-Wesley, Reading, Massachusetts, 1972.

[10] N. Bourbaki, <u>General Topology Part I</u>, Addison-Wesley, Reading, Massacusetts, 1966.

[11] N. Bourbaki, General Topology Part II, Addison-Wesley, Reading, Massachusetts, 1966.

[12] J.R. Boyd, "Axioms that define semi-metric, Moore, and metric spaces", Proc. Amer. Math. Soc. 13 (1962), 482-484.

[13] K.A. Broughan, "A metric characterizing Čech dimension zero", Proc. Amer. Math. Soc. 39 (1973), 437-440.

[14] K.A. Broughan, "Metrization of spaces having Čech dimension zero", Bull. Austral. Math. Soc. 9 (1973), 161-168.

[15] K.A. Broughan and M. Schroder, "Variations on a metric theme", Math. Chronicle 3 (1974), 71-80.

[16] M. Brown, "Semi-metric spaces", Summer Institute on Set Theoretic Topology, Madison, Amer. Math. Soc. (1955), 64-66.

[17] R. Cacciopoli, "Un teorema generale sull'esistenza di elementi uniti in una transformazione funzionale", Rend. Accad. Naz. Lincei 11 (1930), 794-799.

[18] P.J. Collins, "On uniform connection properties", Amer. Math. Monthly 78 (1971), 372-374.

[19] W.W. Comfort, "A survey of cardinal invariants", Gen. Top. App. 1 (1971), 163-199.

[20] C.H. Dowker, "Local dimension of normal spaces", Quart. Jour. of Math. Oxford 6 (1955), 101-120.

[21] J. Dugundji, Topology, Allyn and Bacon, Boston, 1966.

[22] R. Engelking, Outline of General Topology, North Holland, Amsterdam, 1968.

[23] H.C. Enos, "Coarse uniformities on the rationals", Proc. Amer. Math. Soc. 34 (1972), 623-626.

[24] B. Fitzpatrick, Jr., "A note on countable dense homogeneity", Fund. Math. 75 (1972), 33-34.

[25] P. Fletcher and W.F. Lindgren, "Transitive quasi-uniformities", J. Math. Anal. and App. 39 (1972), 397-405.

[26] M.K. Fort, Jr., "Homogeneity of infinite products of manifolds with boundary", Pacific J. Math. 12 (1962), 879-884.

[27] J. de Groot, "On a metric that characterizes dimension", Canad. J. Math. 9 (1957), 511-514.

[28] E. Hille and R.S. Phillips, Functional analysis and semi-groups , Amer. Math. Soc. Colloquium Publications, Vol. 31, revised edition, Providence, R.I. (1957).

[29] O. Hölder, "Die Axiome der Quantität und die Lehre vom Mass", Leipzig Ber., Math.-Phys. Cl. 53 (1901), 1-64.

[30] J. Horváth, Topological Vector Spaces and Distributions Volume I, Addison-Wesley, Reading, Massachusetts, 1966.

[31] I. Kaplansky, "Topological methods in valuation theory", Duke Math. J. 14 (1948), 527-541.

[32] N. Kimura, "On a sum theorem in dimension theory", Proc. Japan Acad. 43 (1967), 98-101.

[33] K. Kuratowski, "Quelques problèmes concernant les espaces métriques non-séparables", Fund. Math. 25 (1935), 534-545.

[34] S. Lang, Algebra, Addison-Wesley, Reading, Mass., 1965

[35] S. Lefschetz, Algebraic Topology, Amer. Math. Soc. Colloquium Pub., Vol. 27, New York, 1942.

[36] N. Levine, "On uniformities generated by equivalence relations", Rend. Circ. Mat. Palermo (2) 18 (1969), 62-70.

[37] N. Levine, "Well chained uniformities", Kyunpook Math. J. 11 (1971), 143-149.

[38] K. Mahler, Introduction to p-adic numbers and their functions, Cambridge University Press, London, 1973.

[39] M.R. Mather, "Paracompactness and partitions of unity", preprint (1964).

[40] E. Michael, "A note on paracompact spaces", Proc. Amer Math. Soc. 4 (1953), 831-838.

[41] K. Morita, "Normal families and dimension theory for metric spaces", Math. Ann. 128 (1954), 350-362.

[42] K. Morita, "On the dimension of normal spaces II", Journ. Math. Soc. Japan 2 (1950), 16-33.

[43] K. Nagami, "Paracompactness and strong screenability", Nagoya Math. J. 8 (1955), 83-88.

[44] L. Narici, "On nonarchimedean Banach algebras", Arch. Math. 19 (1968), 428-435.

[45] L.J. Norman, "A sufficient condition for quasi-metrizability of a topological space", Portugaliae Math. 26 (1967), 207-211.

[46] G. Peano, "Intégration par séries des équations différentielles linéaires", Math. Ann. 32 (1888), 450-456.

[47] A.R. Pears, "On quasi-order spaces, normality and paracompactness", Proc. Lond. Math. Soc. 23 (1971), 428-444.

[48] E. Picard, "Mémoire sur la théorie des équations aux derivées partielles et la méthode des approximations successives", J. Math. (4) 6 (1890), 145-210.

[49] J. Pollard, "On extending homeomorphisms on zero-dimensional spaces", Fund. Math. 67 (1970), 39-48.

[50] V.I. Ponomarev, "Projective spectra and continuous mappings of paracompacta", Mat. Sb. 60 (102) (1963), 89-119; Amer. Math. Soc. Transl. Ser. 2 39 (1964), 133-164.

[51] M. van der Put, "Algèbres de fonctions continues p-adiques", Nederl. Akad. Wetensch. Proc. Ser. A 71 (1968), 401-420.

[52] M.O. Rabin, "Computable algebra, general theory and theory of computable fields", Trans. Amer. Math. Soc. 95 (1960), 341-360.

[53] I.L Reilly, "On quasiuniform spaces and quasipseudo-metrics", Math. Chronicle 1 Pt II (1970), 71-76.

[54] A.P. Robertson and Wendy Robertson, "A note on the completion of a uniform space", Jour. Lond. Math. Soc. 33 (1958), 181-185.

[55] H. Rogers, Theory of Recursive Functions and Effective Computability, McGraw-Hill, New York, 1967.

[56] P. Roy, "Failure of the equivalence of dimension concepts for metric spaces", Bull. Amer. Math. Soc. 68 (1962), 609-613.

[57] P. Roy, "Nonequality of dimensions for metric spaces", Trans. Amer. Math. Soc. 134 (1968), 117-132.

[58] N. Schilkret, "Non-Archimedean Banach algebras", Duke Math. J. 37 (1970), 315-322.

[59] W. Sierpinski, Introduction to General Topology, University of Toronto Press, Toronto, 1934.

[60] M. Sion and G. Zelmer, "On quasi-metrizability", Canad. J. Math. 19 (1967), 1243-1249.

[61] Yu. M. Smirnov, "On strongly paracompact spaces", Izv. Akad. Nauk. SSSR ser. mat. 20 (1956), 253-274 (Russian).

[62] L.A. Steen, "Conjectures and counterexamples in metrization theory", Amer. Math. Mon. 79 (1972), 113-132.

[63] R.A. Stoltenberg, "On quasi-metric spaces", Duke Math. J. 36 (1969), 65-71.

[64] A.H. Stone, "Paracompactness and product spaces", Bull. Amer. Math. Soc. 54 (1948), 977-982.

[65] W.J. Thron, Topological Structures, Holt, Rinehart and Winston, New York, 1966.

[66] H.E. Vaughn, "On the class of metrics defining a metrizable space", Bull. Amer. Math. Soc. 44 (1938), 557-561.

'ol. 342: Algebraic K-Theory II, "Classical" Algebraic K-Theory, ınd Connections with Arithmetic. Edited by H. Bass. XV, 527 ages. 1973.

'ol. 343: Algebraic K-Theory III, Hermitian K-Theory and Geoıetric Applications. Edited by H. Bass. XV, 572 pages. 1973.

'ol. 344: A. S. Troelstra (Editor), Metamathematical Investigation f Intuitionistic Arithmetic and Analysis. XVII, 485 pages. 1973.

ol. 345: Proceedings of a Conference on Operator Theory. dited by P. A. Fillmore. VI, 228 pages. 1973.

ol. 346: Fučík et al., Spectral Analysis of Nonlinear Operators. 287 pages. 1973.

ol. 347: J. M. Boardman and R. M. Vogt, Homotopy Invariant gebraic Structures on Topological Spaces. X, 257 pages. 73.

ol. 348: A. M. Mathai and R. K. Saxena, Generalized Hypereometric Functions with Applications in Statistics and Physical ciences. VII, 314 pages. 1973.

l. 349: Modular Functions of One Variable II. Edited by W. Kuyk d P. Deligne. V, 598 pages. 1973.

l. 350: Modular Functions of One Variable III. Edited by W. ıyk and J.-P. Serre. V, 350 pages. 1973.

l. 351: H. Tachikawa, Quasi-Frobenius Rings and Generalizans. XI, 172 pages. 1973.

. 352: J. D. Fay, Theta Functions on Riemann Surfaces. V, 7 pages. 1973.

. 353: Proceedings of the Conference. on Orders, Group ıgs and Related Topics. Organized by J. S. Hsia, M. L. Madan d T. G. Ralley. X, 224 pages. 1973.

. 354: K. J. Devlin, Aspects of Constructibility. XII, 240 pages. 3.

. 355: M. Sion, A Theory of Semigroup Valued Measures. 140 pages. 1973.

. 356: W. L. J. van der Kallen, Infinitesimally Central Extenns of Chevalley Groups. VII, 147 pages. 1973.

. 357: W. Borho, P. Gabriel und R. Rentschler, Primideale Einhüllenden auflösbarer Lie-Algebren. V, 182 Seiten. 1973.

358: F. L. Williams, Tensor Products of Principal Series resentations. VI, 132 pages. 1973.

359: U. Stammbach, Homology in Group Theory. VIII, 183 es. 1973.

360: W. J. Padgett and R. L. Taylor, Laws of Large Numbers Normed Linear Spaces and Certain Fréchet Spaces. VI, pages. 1973.

361: J. W. Schutz, Foundations of Special Relativity: Kinec Axioms for Minkowski Space-Time. XX, 314 pages. 1973.

362: Proceedings of the Conference on Numerical Solution rdinary Differential Equations. Edited by D. B. Bettis. VIII, 490 es. 1974.

363: Conference on the Numerical Solution of Differential ations. Edited by G. A. Watson. IX, 221 pages. 1974.

364: Proceedings on Infinite Dimensional Holomorphy. ed by T. L. Hayden and T. J. Suffridge. VII, 212 pages. 1974.

365: R. P. Gilbert, Constructive Methods for Elliptic Equa-. VII, 397 pages. 1974.

366: R. Steinberg, Conjugacy Classes in Algebraic Groups es by V. V. Deodhar). VI, 159 pages. 1974.

367: K. Langmann und W. Lütkebohmert, Cousinverteilunund Fortsetzungssätze. VI, 151 Seiten. 1974.

368: R. J. Milgram, Unstable Homotopy from the Stable of View. V, 109 pages. 1974.

369: Victoria Symposium on Nonstandard Analysis. Edited Hurd and P. Loeb. XVIII, 339 pages. 1974.

370: B. Mazur and W. Messing, Universal Extensions and Dimensional Crystalline Cohomology. VII, 134 pages. 1974.

Vol. 371: V. Poenaru, Analyse Différentielle. V, 228 pages. 1974.

Vol. 372: Proceedings of the Second International Conference on the Theory of Groups 1973. Edited by M. F. Newman. VII, 740 pages. 1974.

Vol. 373: A. E. R. Woodcock and T. Poston, A Geometrical Study of the Elementary Catastrophes. V, 257 pages. 1974.

Vol. 374: S. Yamamuro, Differential Calculus in Topological Linear Spaces. IV, 179 pages. 1974.

Vol. 375: Topology Conference. Edited by R. F. Dickman Jr. and P. Fletcher. X, 283 pages 1974.

Vol. 376: I. J. Good and D. B. Osteyee, Information, Weight of Evidence. The Singularity between Probability Measures and Signal Detection. XI, 156 pages. 1974.

Vol. 377: A. M. Fink, Almost Periodic Differential Equations. VIII, 336 pages. 1974.

Vol. 378: TOPO 72 – General Topology and its Applications. Proceedings 1972. Edited by R. A. Alò, R. W. Heath and J. Nagata. XIV, 651 pages. 1974.

Vol. 379: A. Badrikian et S. Chevet, Mesures Cylindriques, Espaces de Wiener et Fonctions Aléatoires Gaussiennes. X, 383 pages. 1974.

Vol. 380: M. Petrich, Rings and Semigroups. VIII, 182 pages. 1974.

Vol. 381: Séminaire de Probabilités VIII. Edité par P. A. Meyer. IX, 354 pages. 1974.

Vol. 382: J. H. van Lint, Combinatorial Theory Seminar Eindhoven University of Technology. VI, 131 pages. 1974.

Vol. 383: Séminaire Bourbaki – vol. 1972/73. Exposés 418–435. IV, 334 pages. 1974.

Vol. 384: Functional Analysis and Applications, Proceedings 1972. Edited by L. Nachbin. V, 270 pages. 1974.

Vol. 385: J. Douglas Jr. and T. Dupont, Collocation Methods for Parabolic Equations in a Single Space Variable (Based on C'-Piecewise-Polynomial Spaces). V, 147 pages. 1974.

Vol. 386: J. Tits, Buildings of Spherical Type and Finite BN-Pairs. X, 299 pages. 1974.

Vol. 387: C. P. Bruter, Eléments de la Théorie des Matroïdes. V, 138 pages. 1974.

Vol. 388: R. L. Lipsman, Group Representations. X, 166 pages. 1974.

Vol. 389: M.-A. Knus et M. Ojanguren, Théorie de la Descente et Algèbres d' Azumaya. IV, 163 pages. 1974.

Vol. 390: P. A. Meyer, P. Priouret et F. Spitzer, Ecole d'Eté de Probabilités de Saint-Flour III – 1973. Edité par A. Badrikian et P.-L. Hennequin. VIII, 189 pages. 1974.

Vol. 391: J. W. Gray, Formal Category Theory: Adjointness for 2-Categories. XII, 282 pages. 1974.

Vol. 392: Géométrie Différentielle, Colloque, Santiago de Compostela, Espagne 1972. Edité par E. Vidal. VI, 225 pages. 1974.

Vol. 393: G. Wassermann, Stability of Unfoldings. IX, 164 pages. 1974.

Vol. 394: W. M. Patterson, 3rd, Iterative Methods for the Solution of a Linear Operator Equation in Hilbert Space – A Survey. III, 183 pages. 1974.

Vol. 395: Numerische Behandlung nichtlinearer Integrodifferential- und Differentialgleichungen. Tagung 1973. Herausgegeben von R. Ansorge und W. Törnig. VII, 313 Seiten. 1974.

Vol. 396: K. H. Hofmann, M. Mislove and A. Stralka, The Pontryagin Duality of Compact O-Dimensional Semilattices and its Applications. XVI, 122 pages. 1974.

Vol. 397: T. Yamada, The Schur Subgroup of the Brauer Group. V, 159 pages. 1974.

Vol. 398: Théories de l'Information, Actes des Rencontres de Marseille-Luminy, 1973. Edité par J. Kampé de Fériet et C.-F. Picard. XII, 201 pages. 1974.

Vol. 399: Functional Analysis and its Applications. Proceedings 1973. Edited by H. G. Garnir, K. R. Unni and J. H. Williamson. II, 584 pages. 1974.

Vol. 400: A Crash Course on Kleinian Groups. Proceedings 1974. Edited by L. Bers and I. Kra. VII, 130 pages. 1974.

Vol. 401: M. F. Atiyah, Elliptic Operators and Compact Groups. V, 93 pages. 1974.

Vol. 402: M. Waldschmidt, Nombres Transcendants. VIII, 277 pages. 1974.

Vol. 403: Combinatorial Mathematics. Proceedings 1972. Edited by D. A. Holton. VIII, 148 pages. 1974.

Vol. 404: Théorie du Potentiel et Analyse Harmonique. Edité par J. Faraut. V, 245 pages. 1974.

Vol. 405: K. J. Devlin and H. Johnsbråten, The Souslin Problem. VIII, 132 pages. 1974.

Vol. 406: Graphs and Combinatorics. Proceedings 1973. Edited by R. A. Bari and F. Harary. VIII, 355 pages. 1974.

Vol. 407: P. Berthelot, Cohomologie Cristalline des Schémas de Caracteristique p > o. II, 604 pages. 1974.

Vol. 408: J. Wermer, Potential Theory. VIII, 146 pages. 1974.

Vol. 409: Fonctions de Plusieurs Variables Complexes, Séminaire François Norguet 1970–1973. XIII, 612 pages. 1974.

Vol. 410: Séminaire Pierre Lelong (Analyse) Année 1972–1973. VI, 181 pages. 1974.

Vol. 411: Hypergraph Seminar. Ohio State University, 1972. Edited by C. Berge and D. Ray-Chaudhuri. IX, 287 pages. 1974.

Vol. 412: Classification of Algebraic Varieties and Compact Complex Manifolds. Proceedings 1974. Edited by H. Popp. V, 333 pages. 1974.

Vol. 413: M. Bruneau, Variation Totale d'une Fonction. XIV, 332 pages. 1974.

Vol. 414: T. Kambayashi, M. Miyanishi and M. Takeuchi, Unipotent Algebraic Groups. VI, 165 pages. 1974.

Vol. 415: Ordinary and Partial Differential Equations. Proceedings 1974. XVII, 447 pages. 1974.

Vol. 416: M. E. Taylor, Pseudo Differential Operators. IV, 155 pages. 1974.

Vol. 417: H. H. Keller, Differential Calculus in Locally Convex Spaces. XVI, 131 pages. 1974.

Vol. 418: Localization in Group Theory and Homotopy Theory and Related Topics. Battelle Seattle 1974 Seminar. Edited by P. J. Hilton. VI, 172 pages 1974.

Vol. 419: Topics in Analysis. Proceedings 1970. Edited by O. E. Lehto, I. S. Louhivaara, and R. H. Nevanlinna. XIII, 392 pages. 1974.

Vol. 420: Category Seminar. Proceedings 1972/73. Edited by G. M. Kelly. VI, 375 pages. 1974.

Vol. 421: V. Poénaru, Groupes Discrets. VI, 216 pages. 1974.

Vol. 422: J.-M. Lemaire, Algèbres Connexes et Homologie des Espaces de Lacets. XIV, 133 pages. 1974.

Vol. 423: S. S. Abhyankar and A. M. Sathaye, Geometric Theory of Algebraic Space Curves. XIV, 302 pages. 1974.

Vol. 424: L. Weiss and J. Wolfowitz, Maximum Probability Estimators and Related Topics. V, 106 pages. 1974.

Vol. 425: P. R. Chernoff and J. E. Marsden, Properties of Infinite Dimensional Hamiltonian Systems. IV, 160 pages. 1974.

Vol. 426: M. L. Silverstein, Symmetric Markov Processes. X, 287 pages. 1974.

Vol. 427: H. Omori, Infinite Dimensional Lie Transformation Groups. XII, 149 pages. 1974.

Vol. 428: Algebraic and Geometrical Methods in Topology, Proceedings 1973. Edited by L. F. McAuley. XI, 280 pages. 1974.

Vol. 429: L. Cohn, Analytic Theory of the Harish-Chandra C-Function. III, 154 pages. 1974.

Vol. 430: Constructive and Computational Methods for Differential and Integral Equations. Proceedings 1974. Edited by D. L. Colton and R. P. Gilbert. VII, 476 pages. 1974.

Vol. 431: Séminaire Bourbaki – vol. 1973/74. Exposés 436–452 IV, 347 pages. 1975.

Vol. 432: R. P. Pflug, Holomorphiegebiete, pseudokonvexe Gebiete und das Levi-Problem. VI, 210 Seiten. 1975.

Vol. 433: W. G. Faris, Self-Adjoint Operators. VII, 115 pages 1975.

Vol. 434: P. Brenner, V. Thomée, and L. B. Wahlbin, Beso Spaces and Applications to Difference Methods for Initial Valu Problems. II, 154 pages. 1975.

Vol. 435: C. F. Dunkl and D. E. Ramirez, Representations of Commutative Semitopological Semigroups. VI, 181 pages. 197

Vol. 436: L. Auslander and R. Tolimieri, Abelian Harmonic Analysi Theta Functions and Function Algebras on a Nilmanifold. V, 9 pages. 1975.

Vol. 437: D. W. Masser, Elliptic Functions and Transcendenc XIV, 143 pages. 1975.

Vol. 438: Geometric Topology. Proceedings 1974. Edited b L. C. Glaser and T. B. Rushing. X, 459 pages. 1975.

Vol. 439: K. Ueno, Classification Theory of Algebraic Varieti and Compact Complex Spaces. XIX, 278 pages. 1975

Vol. 440: R. K. Getoor, Markov Processes: Ray Processes a Right Processes. V, 118 pages. 1975.

Vol. 441: N. Jacobson, PI-Algebras. An Introduction. V, 115 pag 1975.

Vol. 442: C. H. Wilcox, Scattering Theory for the d'Alemb Equation in Exterior Domains. III, 184 pages. 1975.

Vol. 443: M. Lazard, Commutative Formal Groups. II, 236 pag 1975.

Vol. 444: F. van Oystaeyen, Prime Spectra in Non-Commutat Algebra. V, 128 pages. 1975.

Vol. 445: Model Theory and Topoi. Edited by F. W. Lawv C. Maurer, and G. C. Wraith. III, 354 pages. 1975.

Vol. 446: Partial Differential Equations and Related Topi Proceedings 1974. Edited by J. A. Goldstein. IV, 389 pag 1975.

Vol. 447: S. Toledo, Tableau Systems for First Order Num Theory and Certain Higher Order Theories. III, 339 pages. 19

Vol. 448: Spectral Theory and Differential Equations. Proceedi 1974. Edited by W. N. Everitt. XII, 321 pages. 1975.

Vol. 449: Hyperfunctions and Theoretical Physics. Proceedi 1973. Edited by F. Pham. IV, 218 pages. 1975.

Vol. 450: Algebra and Logic. Proceedings 1974. Edited by J Crossley. VIII, 307 pages. 1975.

Vol. 451: Probabilistic Methods in Differential Equations. Proc dings 1974. Edited by M. A. Pinsky. VII, 162 pages. 1975.

Vol. 452: Combinatorial Mathematics III. Proceedings 1 Edited by Anne Penfold Street and W. D. Wallis. IX, 233 pa 1975.

Vol. 453: Logic Colloquium. Symposium on Logic Held at Bos 1972–73. Edited by R. Parikh. IV, 251 pages. 1975.

Vol. 454: J. Hirschfeld and W. H. Wheeler, Forcing, Arithm Division Rings. VII, 266 pages. 1975.

Vol. 455: H. Kraft, Kommutative algebraische Gruppen und R III, 163 Seiten. 1975.

Vol. 456: R. M. Fossum, P. A. Griffith, and I. Reiten, Trivia tensions of Abelian Categories. Homological Algebra of T Extensions of Abelian Categories with Applications to Theory. XI, 122 pages. 1975.